THE SOCIAL HISTORY OF
M^{THE}achine
Gun

New foreword and bibliographical essay by
Edward C. Ezell, Smithsonian Institution

The Johns Hopkins University Press   BALTIMORE

THE SOCIAL HISTORY OF
THE
Machine
Gun

John Ellis

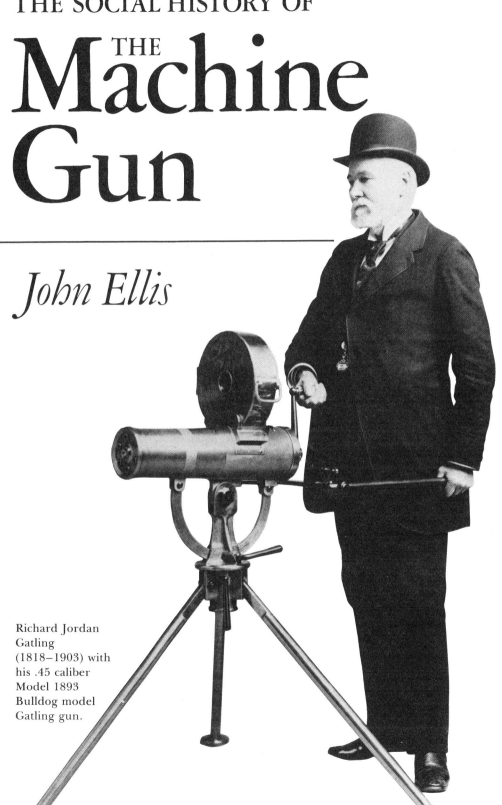

Richard Jordan
Gatling
(1818–1903) with
his .45 caliber
Model 1893
Bulldog model
Gatling gun.

Published by arrangement with Pantheon Books, a division of
Random House, Inc.

Johns Hopkins Paperbacks edition published, 1986
Second printing, 1987

The Johns Hopkins University Press
701 West 40th Street
Baltimore, Maryland 21211

Library of Congress Cataloging-in-Publication Data
Ellis, John, 1945–
    The social history of the machine gun.

    Reprint. Originally published: New York: Pantheon Books, © 1975.
    Bibliography: p.
    Includes index.
    1. Machine-guns—History. 2. Military history, Modern—19th cen-
tury. 3. Military history, Modern—20th century. 4. War and society.
5. Sociology, Military. I. Title.
UF620.A2E38 1986   355.8′2424′09   86-45457
ISBN 0-8018-3358-2 (pbk.)

The illustrations are reproduced by courtesy of the Armed Forces History
Division, National Museum of American History, Smithsonian Institution,
title page and page 27; Griffin and Co., pages 9 and 79; Illustrated News-
papers group, pages 21, 32, 46, 47, 60, 67, 81, 85, 86, 87, and 93; Browing,
page 41; Imperial War Museum, pages 59, 115, 117, and 119; The Tate
Gallery, page 112; Associated Press, pages 154, 156, 169, and 170; Homer
Dickens Foundation, pages 161 and 162; and the U.S. Department of
Defense, the General Electric Company, British Army Public Affairs, and
Fabrique Nationale, Belgium, pages 177–78.

THE SOCIAL HISTORY OF

M<small>THE</small>achine
Gun

Foreword

When John Ellis's *Social History of the Machine Gun* first appeared in print in 1976, it had a provocative impact. It was widely reviewed and discussed in academic and popular publications. Some reviewers suggested that the book was too short to treat the subject properly. Others suggested that Ellis had attempted to attribute too wide a sweep of effects to a single implement of warfare. But nearly all of the people who encountered the book realized that it would provoke thought and discussion. A decade later, *The Social History of the Machine Gun* still seems fresh and relevant. It has helped persuade many readers that military technology and its relationship to our social existence are worthy subjects for reflection. Ellis, as noted in my bibliographical essay at the end of this edition, was not the first writer to tackle the social history of military technology, but he did approach the subject with a flair that caused the reader to reflect upon the mechanization and industrialization of killing. This new edition gives continued circulation to a book that can stimulate students, teachers, military officials, politicians, and lay persons.

Ellis's thesis is straightforward. A product of the age of machines produced in large numbers with interchangeable parts, the rifle caliber machine gun evolved substantially from its introduction in the 1860s until the eve of World War I. In the second half of the nineteenth century, this weapon was seen primarily as an instrument by which small numbers of European soldiers could defeat large masses of native troops in Africa, Asia, and elsewhere. By the time World War I began, machine gun technology had progressed from cumbersome hand-operated, wheeled-carriage mounted, artillerylike weapons to self-actuated, man-portable guns that could be fired at high rates against an advancing enemy from a static position, or fired from one location, then moved to another position, and quickly put into action again.

Hiram Maxim, the inventor of the first self-operated machine gun, called it a "killing machine." In World War I, the machine gun, instead of being used to keep native populations under control, was used to kill thousands of Europe's youth. The potential for this latter application of the machine gun—for causing the "lost generation"—was not perceived by most European military leaders when they first adopted the machine gun to help them consolidate their nation's empires.

John Ellis sets out to establish that "guns, like everything else, have their social history." He explains that the machine gun was adopted as an instrument of war only partly because it was a more effective piece of technology. The goals of the military officer class, the amoral (as opposed to immoral) problem-solving concerns of the inventors, the profit motivation of the arms makers, and the rush to establish empires in Africa and elsewhere all influenced the decision to develop, adopt, purchase, and manufacture the machine gun.

Ellis thus argues an important point: "The history of technology is part and parcel of social history in general." Technology cannot be studied in isolation. An implicit element of his book is the belief that military technology is an acceptable subject for study. Ellis makes it clear that an examination of the military and its equipment does not convey approval of either. Many "liberal" historians have shied away from the analysis of military technology and its impact on our lives, because they find the subject distasteful. Ellis is certainly neither promilitary nor prowar, but he has laid out in clear and useful terms how the machine gun came about, why it was adopted, and what some of the consequences of its use have been.

Millions of automatic weapons—machine guns, automatic rifles, and submachine guns—have been made and deployed with the world's armies since Hiram Maxim introduced his first self-actuated machine gun in 1886. One example of the production levels illustrates this point. Since its debut in 1947, some thirty to fifty million Kalashnikov AK assault rifles have been manufactured by the Soviets and their allies. Like the Maxim, the Kalashnikov is more than a weapon; it has become a symbol. The former represented the power of the imperial armies, while the AK has become an icon for many of the anti-establishment insurgent, freedom fighter, and terrorist organizations that exist today. The effectiveness of these weapons as symbols comes from their efficiency as machines of death. And it should be noted that most of the colonial regimes in Africa and Asia ceased to exist when the indigenous population "got" the machine gun in numbers sufficient to overwhelm, or threaten to overwhelm, the power of the imperial state. Mao Tse Tung correctly noted that "power comes from the muzzle of the gun." Neither Hiram Maxim nor Mikhail Timofeyevich Kalashnikov would argue with that tenet.

The value of Ellis's *Social History of the Machine Gun* goes far beyond the case study. By stimulating students of history to look at the machine gun, Ellis has established a pattern that can be applied to investigation of the military, social, and policy issues raised by current programs to develop new military technology. From AK47s in the hands of terrorists, killing innocent civilians in airports and restaurants, to multibillion-dollar proposals for Star Wars (Strategic Defense Initiative) defensive shields, military and civilian leaders still talk simultaneously, alternatively, and confusingly of the benefits and the threats of military technology.

As one reviewer noted: "Strangelovingly, the machine gun was invented by an American, Richard Gatling, whose familiar hope it was that wars would be prevented when people who fought them realized the heavy toll his weapon would exact. It was that kind of logic that started the arms race and made Gatling a wealthy man." The arms race and profits from weapons production continue to be facts of life. Ellis's *Social History of the Machine Gun* is, and will continue to be, a good introduction for the thinking person who wants to understand more of the social history of military technology, past and present.

Edward Clinton Ezell
Division of Armed Forces History
National Museum of American History
Smithsonian Institution
Washington, D.C.

Acknowledgements

I am most grateful to the following for their permission to reproduce extracts from the various works named below:

Mr.G.T.Sassoon for portions of 'The Kiss' and 'The Redeemer' by Siegfried Sassoon.

Gerald Duckworth Ltd. for a portion of 'The Modern Traveller' by Hilaire Belloc.

Peter Newbolt for a portion of 'Vitai Lampada' from *Poems Old and New* by Sir Henry Newbolt.

Mrs.G.Bambridge and Macmillan and Co. for portions of 'Pharoah and the Sergeant' from *The Definitive Edition of Kipling's Verse*, 'The White Man's Burden' and 'The Lesson' from *The Five Nations*, all by Rudyard Kipling.

I am also grateful to the following for their help in finding and their permission to use certain of the illustrations in this book:

The Browning Company of Morgan (Utah)

Mr.William J.Helmer, author of *The Gun That Made the Twenties Roar*.

Illustrated Newspapers Ltd. and Mr.H.E.Bray of the Copyright and Archives Department.

Associated Press and Mr. Ray Pereira of the Photograph Library.

Contents

I *New Ways of Death*

Fear not, my Friends, this Terrible Machine,
They're only Wounded that have Shares therein.

Anonymous pamphleteer, 1718.

Machine guns are now commonplace. The indispensable aid they offer to soldiers, policemen and terrorists is taken for granted. Yet this acceptance has not come easily. The reasons for this are numerous, but most of them are much more than a simple evaluation of the machine gun's technical merits. The following pages will show that the general aspirations and prejudices of particular social groups are just as important for the history of military technology as are straightforward problems of technical efficiency. Guns, like everything else, have their social history. In this book it will be seen that the anachronistic ideals of the European officer class, the messianic nature of nineteenth-century capitalism, the imperialist drive into Africa and elsewhere, and the racialist assumptions that underpinned it, were more important to the history of the machine gun than any bald assessment of its mechanical efficacy. The history of technology is part and parcel of social history in general. The same is equally true of military history, far too long regarded as a simple matter of tactics and technical differentials. Military history too can only be understood against the wider social background. For as soon as one begins to discuss war and military organisation without due regard to the whole social

9

process, one is in danger of coming to regard it as a constant, an inevitable feature of international behaviour. In other words, if one is unable to regard war as a function of particular forms of social and political organisation and particular stages of historical development, one will not be able to conceive of even the possibility of a world without war.

Before Their Time

Only a hundred years ago machine guns were generally regarded as being little more than mechanical gimmicks, of no real value on the conventional battlefield. In the beginning this lack of interest in the potential of sustained automatic fire was quite understandable. It was always theoretically possible to conceive of a gun that would somehow spew out vast numbers of bullets or whatever in a very short time. But for hundreds of years men lacked the technological expertise with which to translate such visions into reality. There were no metals that could withstand the pressures of regular mass or sustained fire. Nor were manufacturing techniques sufficiently well advanced to enable individual craftsmen to work to the fractional tolerances demanded for every part of such a complex gun.

The first efforts in the field of increasing the firepower of an individual piece were limited to trying to multiply the volume of each discharge rather than increasing a gun's rate of fire. Since the very introduction of gunpowder, in fact, there had always been those who were keen to make such an improvement. The earliest efforts in this direction were the *ribauldequins*, or 'organ guns'. These ponderous contraptions consisted of varying numbers of barrels laid out in parallel layers, each of which could be ignited almost simultaneously by one or two gunners. The first mention of such a weapon was in 1339, and they seem to have soon achieved considerable popularity. In 1382 the army at Ghent had two hundred of these guns in the field, whilst in 1411 the Burgundian army is reputed to have had some two thousand of them. In 1457 the Venetian general, Colleoni, employed organ guns at the battle of Picardini as a mobile auxiliary to his armoured cavalry. They were also used against the French by Pedro Navarro, who placed thirty carts of multi-barrel guns in front of his infantry. These devices had some impact on the French, and Louis XII (1498 – 1515) is said to have had at his disposal at least one gun of fifty barrels, so arranged that they could all be fired simultaneously. But such guns had severe limitations. Even assuming that all barrels fired without mishap, the initial effect of the volley was severely dissipated by the inordinate length of time it took to muzzle-load each individual barrel. The advantages gained by the ability to deliver concentrated fire were nullified by the inability to maintain a sustained fire. Similarly, the guns were so unwieldy that it was almost impossible to manoeuvre them to positions on the battlefield where they might be most needed.

For hundreds of years in fact gunnery was a very crude and unreliable science. Up to the sixteenth century, for example, the actual firing of a gun depended upon the bringing of some form of fire, usually a slow match, into contact with the powder in the firing pan. After the sixteenth century the process was streamlined somewhat and made a little more reliable. From this date the initial ignition was achieved by striking together flint and roughened steel to produce a spark. Such flintlocks were the standard infantry arm until the invention of the percussion cap in 1807. In neither period was it technically feasible to produce a reliable gun that could produce either a concentrated or a sustained fire. In both cases such a gun would be bedevilled by the same problem of the prolonged loading period that had so detracted from the value of the organ guns. Similarly, the chronic unreliability of both loading methods would not guarantee any multi-barrel gun much more than a fifty per cent fire rate; whilst the combination of this same unreliability, the instability of many charges of gunpowder packed into one piece, and the possibility of serious flaws in the barrels due to crude casting techniques, could make any such gun a potential death trap for those called upon to fire it.

Even so, the very idea of being able to produce a gun that had a significantly superior firepower occasionally drove an inventor to the drawing board. Either from ignorance or blind faith each went to work thinking that for him at least the constraints of an as yet inadequate technology would count for nothing. One such visionary, in the sixteenth century; came to see Sir Francis Walsingham, then the Secretary of State. With him he had a letter that proclaimed that its bearer, a German, 'among other excellent qualities which he hath, can make an harquebus that shall contain balls or pellets of lead, all of which shall go off one after another, having once given fire, so that with one harquebus one may kill ten thieves or other enemies without recharging it.' In 1626 Charles I granted a patent to a certain William Drummond in Scotland. He had described his invention as being 'a machine in which a number of musket barrels are fastened together in such a manner as to allow one man to take the place of a hundred musketeers in battle.' The machine was to consist of fifty barrels mounted in a circular fashion. Not only would the gun be able to fire fifty rounds at

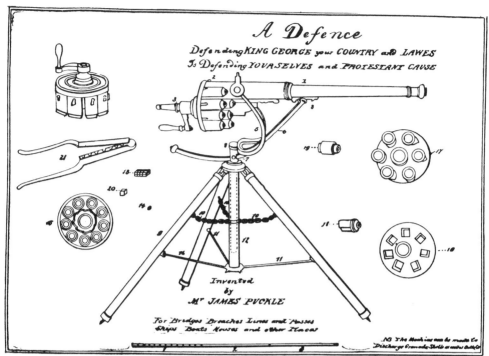

A Defence

Defending KING GEORGE your COUNTRY and LAWES
Is Defending YOUR SELVES and PROTESTANT CAUSE

Invented
by
Mr JAMES PUCKLE

For Bridges Breaches Lines and Passes
Ships Boats Houses and other Places

NB The Machine can be made to
Discharge Granado Shells at once both

No. 1 The Barrel of the Gun
2 The Sett of Chambers Charg'd put on ready for Firing
3 The Screw upon which every Sett of Chambers play off and on
4 a Sett of Chambers ready charged to be Slip'd on when the first Sett are pull'd off to be recharg'd
5 The Crane to rise fall and Turn the Gun round
6 The Curb to Level and fix the Guns
7 The Screw to rise and fall it
8 The Screw to take out the Crane when the Gun with the Trepeid is to be folded up
9 The Trepied whereon it plays
10 The Chain to prevent the Trepieds extending too far out
11 The hooks to fix the Trepied and Unhook when the same is folded up in order to be carried with the Gun upon a Man's Shoulder
12 The Tube wherein the Pivot of the Crane turns
13 a Charge of Twenty Square Bullets
14 a single Bullet
15 The front of the Chambers of a Gun for a Boat
16 The plate of the Chambers of the Gun for a Ship shooting Square Bullets against Turks
17 For Round Bullets against Christians
18 a Single Square Chamber
19 a Single round Chamber
20 a single Bullet for a Boat
21 The Mould for Casting Single Bullets

Whereas our Soveraign Lord King George by his Letters pattents bearing date the Fifteenth day of May in the Fourth Year of his Majesties Reign was Graciously pleas'd to Give & Grant unto me James Puckle of London Gent my Exers Admors & Assignes the Sole priviledge & Authority to Make Exercise Work & use a Portable Gun or Machine (by me lately Invented) called a Defence in that part of his Majesties Kingdom of Great Brittain called England his Dominion of Wales Town of Barwick upon Tweed and his Majesties Kingdom of Ireland in such manner & with such Materials as Should be ascertain'd to be the sd New Invention by writing under my Hand Seal and Inrolled in the High court of chancery within Three Calendar Months from the date of the sd pattent as in & by his Majs: Letters Pattents Relacon being thereunto had Doth & may amongst other things more fully & at large appear NOW I the said James Puckle Do hereby Declare that the Materials whereof the sd Machine is Made are Steel Iron & Brass and that the Trepied whereon it Stands is Wood & Iron And that in the above print (to which I hereby Refer) the said Gun or Machine by me Invented is Delineated & Described July the 25th 1718./

Ja Puckle

Puckle's original patent

12

once, but Drummond also envisaged it being able to fire several volleys before it needed reloading. For each barrel was to be loaded with one charge after another, each aligned with touch holes that ran all the way up the barrels. Each barrel was to be ignited again and again by moving up adjustable fuse-holding devices until they were in line with the touch-holes.

In 1663 there occurred a very remarkable theoretical breakthrough that anticipated by over two centuries one of the basic principles of later machine gun development. A man named Palmer presented a paper to the Royal Society in which he explored the possibilities of utilising the force of the recoil and the gases escaping along the barrel to load, discharge and reload the weapon. As he described it, it was a 'piece to shoot as fast as it could and yet be stopped at pleasure, and wherein the motion of the fire and the bullet within was made to charge the piece with powder and bullet, to prime it and to pull back the cock.' Unfortunately, in technical terms, Palmer was about two hundred years ahead of his time. Because the chambers of any gun at that time could not be properly bored, it would be impossible to effectively trap all the escaping gases. And until cartridges were made of a solid-drawn brass or copper case it was impossible to ensure that they were properly ejected. There is in fact no record of such a gun ever having been built. And a man so far ahead of his time would have been lucky if he managed to retain his membership of the Royal Society.

The next authenticated invention of any importance is described in a patent of May 1718, granted to one James Puckle. Therein it is recorded that: 'James Puckle of the City of London, Gentleman, hath at great expense invent "A Portable Gun or Machine called a Defence, that discharges so often and so many bullets, and can be so quickly loaded as renders it next to impossible to carry any ship by boarding." ' The gun basically consisted of a barrel and a revolving chamber, each chamber being discharged at it was brought into alignment with the barrel. If several loaded sets of chambers were kept on hand a fairly sustained rate of fire could have been achieved. However, though Puckle's invention is of considerable interest as one of the first detailed examinations of the revolving chamber principle, it seems unlikely that he himself had much interest in merely advancing the cause of science. Puckle's own drawing on the patent is boldly headed with the following description of his 'Defence':

Defending KING GEORGE your COUNTRY and LAWES

in Defending YOURSELVES and PROTESTANT
CAUSE

In his eagerness to defend his religious orthodoxy and
patriotic zeal Puckle then seems to get rather carried away,
for he goes on to fly in the face of technical possibilities. Thus
his detachable chambers were to be of two types, the one
containing round bullets to be fired against Christians and
the other square bullets for use against the heathen Turks.
Even were it remotely possible to fire square bullets in the
first place, Puckle's design makes no provision for any
change of barrel when switching from one type of ammunition to the other. A satirist of the time made a suitably
scathing attack on Puckle's motives and technical ability:

> A rare invention to Destroy the Crowd
> Of Fools at home, instead of Foes abroad:
> Fear not, my Friends, this Terrible Machine,
> They're only Wounded that have Shares therein.

If Palmer's paper offered a fascinating preview of the
technical aspects of the development of automatic fire, Puckle's promotional efforts had much in common with the dubious motives and methods of those later figures who tried to
sell the now perfected machine gun to a sceptical market.

Even when advances in manufacturing technique had
enabled men like Gatling, Maxim or Nordenfelt to produce
reliable automatic weapons they found that the attitude of
the world at large, particularly the military, had not
advanced much beyond the derisive strictures of Puckle's
anonymous critic. A few guns were brought by various countries during the nineteenth century, but the market was very
limited. The manufacturers and their representatives were
obliged to use decidedly shady means to ensure themselves
at least a slice of the market. For most of them the idea of
commercial success was of overriding importance, and they
made few concessions to either patriotism or normal business ethics. The most cynical and blatant of these merchants
of death was Zaharoff, who worked first for Nordenfelt and
then for the amalgamated firm of Maxim-Nordenfelt. His
early career was devoted to denigrating and actually
sabotaging any rival machine guns, notably the Maxim.
Later when faced with the prospect of actually having to sell
the gun he had so assiduously slandered, he resorted to
bribery as his most reliable technique. On top of this
Zaharoff was one of the principal influences behind the late
nineteenth-century arms race, in which carefully planted
rumours and 'leaks' about other nation's military capacity
forced governments to buy increasing numbers of weapons
to keep ahead of their 'rivals'.

Machine gun development, then, revealed many of the most unattractive aspects of the brash and often brutal spirit of nineteenth-century capitalism. The inventors, complacently proud of having overwhelmed the technical difficulties inherent in producing reliable automatic weapons, saw no reason why they should not sell as many as possible and reap the rewards of their ingenuity. Nor is the relationship between machine gun development and capitalist methods merely limited to the clamour of the market place. One of the very first benefits that men saw in this ability to bring massive firepower to bear from just one gun was its possible use in the war between capital and labour. In America the machine gun was for long enough a military non-starter. But it soon found its civilian adherents and Gatlings, Brownings, Lewises and Thompsons all served with state National Guards or with gangs of company vigilantes who all realised that here was a stunningly economical means of deterring the discontented workers or, if the worst came to the worst, of killing a few as a warning to the rest. Even in the 1930s one harassed American entrepreneur remarked that: 'You can't run a mining company without a few tommy guns.' And in the preceding decades machine guns were invaluable to the businessmen of Colorado and California, Chicago and Philadelphia.

Yet for years such weapons had no attraction for the military, the only consumer capable of allowing machine gun production to get off the ground. At first their lack of interest was quite excusable. Palmer's ideas, for example, were clearly only of theoretical interest. Puckle's machine gun did not even make sense on paper. Even in the nineteenth century many of the ideas that were thrown up were often somewhat bizarre. In 1834, for example, the Danish inventor and military gunmaker, N.J.Løbnitz, produced a full-scale air machine gun. It was tested by a Danish military commission and found to be capable of a maximum fire rate of eighty shots per minute. Naturally the muzzle velocity was found to be somewhat less than that of conven-

Organ Gun

tional firearms, but even so the gun was proved able, at a range of 250 feet, to send a ball through a one-inch pine board. What actually rendered the gun completely worthless from a military point of view was that the vertical pump that fired it was driven by two enormous flywheels, almost six feet in diameter. Given that this cumbrous mechanism would in turn have to be mounted on an artillery carriage, it was not surprising that the gun was swiftly rejected as a feasible military weapon.

In 1854 Sir Henry Bessemer patented a self-acting breech-loading gun that utilised steam to operate the feeding, locking and firing of the piece. Luckily he also patented what is known as the 'Bessemer process' for making steel, and went on to make his fortune. Had he been forced to persevere in the field of armaments the British military might well have been reduced to an army of stokers.

But in 1862 Gatling produced a crank-operated gun that was soon perfected and was quite capable of producing a steady stream of fire at 200 rounds per minute. In 1884 Hiram Maxim demonstrated a perfectly reliable machine gun in which the initial pulling of the trigger made the gun fire completely automatically until the trigger was released. In 1892, in America, William Browning developed his own fully automatic gun, in which the mechanism was operated by the pressure of the muzzle gases. All of these versions of the machine gun were well-designed, relatively easy to mass produce and fairly reliable under battlefield conditions. They all offered their users a quite staggering increase in firepower. As Gatling said of his own piece: 'It bears the same relation to other firearms that McCormack's Reaper does to the sickle, or the sewing machine to the common needle.' But the military establishments of Europe and America did not wish to face up to this fact, and were, in fact, incapable of doing so. For the machine gun was a product of the Industrial Revolution, of the fundamental changes in manufacturing and financial techniques that had gathered pace during the nineteenth century. To the proponents of this massive technological leap the machine was the answer to everything. For them even killing could be mechanised and made more efficient. But the various armies remained outside this school of thought. The bulk of their officers came from those very landowning classes that had been left behind by the Industrial Revolution. They tried to make the army a last bastion of the attitudes and the life-style that had characterised the pre-industrial world. Because of their rigid hierarchical structures and the fact that all promotions had to be sanctioned from above they were able, for decade after decade, to minimise the impact of the new faith in science and the machine. In 1840 the eleven-pound, six-foot musket was still the standard battlefield weapon. And not even the widespread introduction of the rifle made a very significant difference to the traditional military modes of thought. Even in 1914 most professional soldiers still saw the rifle and bayonet, basically a shortened pike, as the ultimate weapon. In essence their tactical scheme of things had altered little since Gustavus Adolphus, Frederick the Great or Napoleon.

When faced with the machine gun and the attendant necessity to rethink all the old orthodoxies about the primacy of the final infantry charge, such soldiers either did not understand the significance of the new weapon at all, or tried to ignore it, dimly aware that is spelled the end of their own conception of war. It would be almost impossible to over-emphasise the resilience of such a myopic outlook amongst the military leaders of the nineteenth century. For them all the progress of the preceding years merely meant that the standard military weapons, the cannon and the musket, became slightly more efficient. Ranges were longer, rates of fire quicker, muzzle velocities higher, but basically, for them, nothing had changed. The bayonet push and the cavalry charge were still the determining factors on the battlefield. Even in 1926 Field-Marshal Haig could assert that 'aeroplanes and tanks . . . are only accessories to the man and the horse, and I feel sure that as time goes on you will find just as much use for the horse . . . as you have ever done in the past.'

Clearly the machine gun, more than anything else, was a dire threat to such assumptions about the nature of war. The officer corps of the nineteenth century clung on to their old beliefs in the centrality of man and the decisiveness of personal courage and individual endeavour. Machines had brought with them industrialisation and the destruction of the traditional social order; they must not be allowed to undermine the old certainties of the battlefield – the glorious charge and the opportunities for individual heroism. The machine gun threatened to do just this. Its phenomenal firepower could render such charges quite futile. It negated all the old human virtues – pluck, fortitude, patriotism, honour – and made them as nothing in the face of a deadly stream of bullets, a quite unassailable mechanical barrier. For the old-style gentleman officers such an impersonal yet utterly decisive baulk was unacceptable. So they tried to ignore it. For them the machine gun was anathema, and even when their governments bought them out of curiosity, or because their enemies did, they almost totally ignored them. The behaviour of certain commanders during manoeuvres just before the First World War perfectly summed up the whole military attitude to the new automatic weapons. When asked by keen young subalterns what they should do with the machine guns they replied: 'Take the damned things to a flank and hide them!'

In England and other countries they remained well hidden until the very outbreak of the First World War. Yet in one part of the world the deadliness of the machine gun had already been made all too clear. In Africa small parties of Europeans, soldiers and armed settlers often had to face the resistance of large numbers of poorly armed natives. The odds were so in favour of the natives that the white men were obliged to adopt all weapons that would help to maximise their firepower. The machine gun made a fine addition to breech-loading rifles. In all parts of the continent, against Zulus, Dervishes, Hereros, Matabele and many other peoples, Gatlings, Gardners and Maxims scythed down anyone who dared to stand in the way of the imperialist advance. Such weapons were absolutely crucial in allowing the Europeans to hang on to their tiny beachheads and give themselves a breathing space for further expansion. Without the handful of machine guns the British South Africa Company might have lost Rhodesia; Lugard might have been driven out of Uganda and the Germans out of Tanganyika. Without Hiram Maxim much of subsequent world history might have been very different. As Hilaire Belloc put it:

Thank God that we have got
The Maxim gun, and they have not.

But these imperialist sideshows, devastating as they were for anybody with the wit to see it, made little impression upon the military *élites* at home. Europeans, particularly the British, were too concerned with trumpeting the virtues of their small squares of heroes to admit that much of the credit for these sickeningly total victories should go to the machine guns. Where was the glory, where was the vicarious excitement for the readers back home, if one told the truth about totally superior firepower? One couldn't pin a medal on a weapon. And one couldn't demonstrate the manifest superiority of the British race by admitting that it was a mere piece of hardware that had swung the balance.

In truth, machine guns had come a long way since the days of Palmer and Puckle, Løbnitz and Bessemer. Neither these dabblers, nor even Gatling and Maxim imagined what a terrible weapon they had created, how greatly they had increased man's capacity to wipe himself off the face of the earth. The imperialist experience had revealed the truth, but no one wanted to face up to it. Anyway, how could military campaigns against a few scruffy 'Kaffirs' teach Europeans anything about the shape of future war on their continent? So the machine gun remained right at the bottom of the procurement lists. It seemed best just to have a few in case one's opponent had them, but there was certainly no point in pretending that they might make one blind bit of difference.

Unfortunately, even these few, in the hands of the German, British and French armies of 1914, *did* make a difference. They gave an overwhelming advantage to the defensive and made a complete nonsense of both sides' visions of what a major European conflict would be like. They, together with adroitly handled rifles, played a part in halting the initial German thrust. They were even more instrumental in bringing the Allies to a grinding halt when they in turn attempted to drive the Germans back. Neither side could make any headway because of the terrible sheet of fire that even a relatively thinly held infantry position was able to produce. Each side was forced to dig holes in the ground to protect themselves from this fire. They remained in these holes for a further four years. Whenever either side attempted to push forward, the Germans at Verdun, the French in Champagne, the British on the Somme, a few well-sited machine guns literally swept the attacking infantry away, causing unparalleled casualties in a matter of minutes. The machine gun had finally come into its own.

But even now there were many who were not prepared to learn the lesson. In one sense the 1914 – 18 War represented the culmination of years of industrial progress. This progress had enabled nations to throw larger and larger armies into the field and to develop more and more sophisticated weapons. But the generals had chosen to ignore the implications of these developments. Even as the war progressed they continued to ignore them. For a Haig or a Joffre it was just not conceivable that mere guns could hold up an attack pressed home with sufficient vigour or *élan*. So for four years the ordinary soldier suffered, trapped between the contradictory forces of the logic of technological advance and the anachronistic conceptions of the professional military mind. Either they rotted in the mud or they were sent over the top to give the machine gunners some not very demanding target practice.

Puckle had thought that his gun might be useful for raking:

   ... Bridges, Breaches, Lines and Passes,
   Ships, Boats, Houses, and other Places.

Sir Francis Walsingham's German visitor had merely

thought that his gun might help to kill 'ten thieves or other enemies'. Gatling had even supposed that such a weapon would 'supersede the necessity of large armies and consequently exposure to battle and disease be greatly diminished'. As it turned out the machine gun became a decisive battlefield weapon, capable of mowing down troops in their hundreds and occasioning the need for ever larger drafts of men. It shaped the course of one of the greatest wars in history, and was responsible for the slaughter of a whole generation of young Europeans. That slaughter left a mark that has affected the history of the whole world.

These, then, are the kinds of issues that will be dealt with in this 'social history' of the machine gun. It will in no way be simply a description of rates of fire, or muzzle velocities, blow-back actions or 'number two stoppages'. This is rather a study of what the history of the machine gun tells us about society at the time, and how the nature of that society helps us to explain the greed, hypocrisy, callousness and sheer bigotry that are so much a feature of the story of automatic fire.

II *Industrialised War*

'You Yankees beat all creation. There seems
to be no limit to what you are able to do.'

Sir Garnet Wolseley to Hiram Maxim.

During the American Civil War (1861–65) workable
machine guns made their first appearance, and from then on
they developed rapidly. As most of this development was in
the United States. it is on this country that the following
discussion will focus. By 1860 the United States, after
Britain, formed the second largest manufacturing nation in the
world, and had far outstripped any other country in the
development of machines to do jobs previously undertaken
by skilled workers, and in the grouping together of such
machines within a single factory. There are several reasons
for this engineering pre-eminence. Most importantly,
perhaps, America, particularly in the early years of the
nineteenth century, was acutely short of manpower. To
attract men to work in the new factories it was necessary to
pay high wages. If prices were not to be unrealistically high it
was necessary that the productivity of this well-paid labour
force be itself high. Thus machines and rationalised, cen-
tralised production units were introduced to multiply the
productivity of the individual worker. Secondly, it is a nota-
ble fact that few of the machines used in the early years were

actually of American origin, most of them being based upon designs or originals pirated from Europe. But these machines were first properly exploited in America because the country completely lacked a well-organised class of handworkers who might look upon mechanisation as a threat to their traditional way of life. In fact there were few workers at all who would have been capable of manning any large-scale production efforts based upon human skill rather than mechanical efficiency. One of the pioneers of mechanical, mass-production methods, Eli Whitney, gave a succinct definition of his reasons for adopting such a mode of manufacture. The purpose was 'to substitute correct and effective operations of machinery for that skill of the artist which is acquired only by long practice and experience; a species of skill which is not possessed in this country to any considerable extent.' Thus there was produced in America 'a new way, not of making things, but of making machines that make things. It was a simple, but a far-reaching change, not feasible in a Europe rich in traditions, institutions and vested skills.'[1]

Thus did the Americans, long before anyone else, find themselves pushed into a dependence upon machine-based industry. Moreover, the very manufacture of the machines themselves, and the necessity to think in terms of mechanical capabilities rather than merely the limits of human skill, threw up a whole breed of experts concerned solely with the creation of better machines. Out of this grew the first significant machine tool industry, and a new interest in the properties of various metals for cutting and grinding purposes, and a new concern with machines that could always reproduce a part of almost exactly identical dimensions. In America, in the nineteenth century, 'machines, not men, became specialised.'[2]

There is one other feature of this reliance upon machines that is of considerable importance. From the very beginning it was always very closely associated with the manufacture of small arms. In some ways this is perfectly understandable in that the army in any country always has a demand for large numbers of identical pieces of equipment, be they firearms, uniforms or accoutrements. But America was even so a special case. In Europe military requirements, especially those for guns, had always been met by the combined efforts of numerous small gunsmiths. Because of the static nature of military technology throughout the sixteenth, seventeenth and eighteenth centuries any army could safely rely upon slowly stockpiling the necessary guns rather than putting in sudden large orders. In America, however, the immemorial practice of the handcrafted gun did not exist and, furthermore, there were simply not enough gunsmiths around to satisfy the demands of the army. These demands in the first years of the nineteenth century were made particularly pressing because 'most of the muskets with which the Americans had won their Revolution fifteen years before had been made in France or elsewhere in Europe . . . Since military firearms had not been manufactured in quantity in America . . . the country was, in effect, unarmed.'[3] The only solution was to turn the problem over to the new engineers.

The great pioneer in the field was Whitney who, in 1798, his country being faced with the prospect of a war with France for which it was totally unprepared, signed a contract with the government pledging the production of ten thousand muskets within the next twenty-eight months. In fact it took him ten years to actually complete the contract. But what is of importance here is not that Whitney fell lamentably behind the delivery date, but that even at this early stage, in the United States, the links between armaments manufacture and machine tooling had already been established.

So far there are discernible three main reasons for the fact that a workable machine gun first appeared in the United States at the time it did. Firstly the peculiarities of American society in the first years of the nineteenth century forced those that wished to build up her productive capacity to rely upon machinery and streamlined production processes. This brought with it a new expertise and interest in the manufacture of ever more complex machinery, and a concern with more durable metals and increasingly exacting standards of accuracy. Secondly, the development of these mechanical skills was always very closely connected with the progress of the American small arms industry. It is not too surprising, therefore, that it should be an American who first solved the problem of making a reliable automatic gun. Thirdly, on a more general level, this dependence upon machinery created a new faith in the unlimited potential of machines, and the belief that anything could be made into an automatic process if only one applied oneself sufficiently diligently. Thus, if it is only natural that an American should have the tools and the know-how to make the first machine gun, it is equally logical that it should be someone from that country who would actually want to turn killing into a matter of turning a crank or depressing a button.

These, then, are the general preconditions for the development of an adequate automatic weapon in the United States. But there is still one more quite specific event which helped to ensure that such a weapon actually saw the light of day. This was the American Civil War. All authorities are agreed that the Civil War was the first truly modern war, in which the effects of the new technology first made themselves apparent. Walter Millis has summed up the main points admirably:

> Once war was joined . . . technology was to intensify the scope, deadliness, and universality of the struggle with a speed that would previously have been impossible. The railroad and river steamboat enabled both sides to mobilise, supply and deploy armies of unprecedented size with an unprecedented promptness . . . In 1861 the Confederate Congress began by voting, two days after Lincoln's inauguration, an army of 100,000 volunteers. It is said to have had one third of them organised and under arms in little over a month . . . Lincoln, who had started by calling for 75,000 militia in mid-April, is said to have had something like 250,000 men under arms by July . . . When in the immediately following days after Bull Run both sides took serious measure of the struggle lying before them, the Northern Congress voted an army of 500,000 men and the Southerners one of 400,000 . . .
>
> Begun on this scale the war was to be waged with a mounting intensity for which, again, the new . . . technology was in large part responsible. It is estimated that from first to last about 900,000 individuals served in the Confederate armies and about 1,500,000 in those of the North. In providing these hosts with weapons, ammunition, accoutrements and clothing, as well as with transportation and supply, industry on both sides was to perform remarkable feats.[4]

The point about the huge numbers of men involved is the central one here. All civil wars tend to be very bloody affairs because each side is fighting for its sheer survival, knowing that defeat will involve the most terrible reprisals. This tendency was accentuated in the American Civil War because it was the first war in which both sides were able to effectively mobilise the potential, in terms of *materiel* and manpower. In previous wars any state's military effort was severely limited by financial, administrative and technological inadequacies. Only a minute proportion of the military age-group could actually be put on the battlefield, and there were even fewer reserves to back them up. Between 1861 and 1865, on the other hand, the two sides were pitting their whole potential against each other. It was not simply a question of killing an adequate proportion of men on a given battlefield, but rather one of having the capacity to fight off, month after month, year after year, the fresh levies thrown in by one's opponent. Soldiers were now much more expendable, as individuals, than they had ever been before. A new emphasis was placed upon the material ability to kill as many men as possible. What one might call the concept of 'overkill' was first introduced into warfare. The days were now gone when it was sufficient to win one big battle to win a war. One now had to fight battle after battle, and in each of them it was vital to kill as many men as possible. Of course the full implications of all this were not really felt until 1915 and 1916, in Europe, but one can nevertheless perceive some of the horrors of industrialised warfare on the battlefields of the American Civil War. It was no coincidence that the machine gun, for many years the ultimate weapon of mass destruction, should make its first appearance during this particular conflict.

The Gatling Gun

The first recorded sale of a machine gun was by J.D.Mills, who was the sales representative for the Ager Gun, or Union Repeating Gun as it is sometimes known. Mills optimistically referred to the gun as 'an Army in six feet square', and managed to arrange a demonstration before President Lincoln. Lincoln was very enthusiastic but his Chief of Ordnance, Colonel J.W.Ripley, an inveterate standardiser, contemptuous of all new inventions, was completely opposed to its adoption. But Mills persevered, and on 16 October 1861 he made the historic sale. His powers of persuasion must have been considerable for Lincoln agreed to buy all the ten guns that were then available at the very high price of $1,300 each. Later in the year General McClellan ordered another fifty guns, though he managed to beat the price down to only $735. This was by far the biggest order Mills ever managed to get, though he did seem to regain his poise as regards the actual cash value of his guns. In 1861 General Butler was prevailed upon to buy two of them for $1,300 each, and in the following year General Fremont laid out $1,500 for each of two guns. But their performance on the battlefield never seemed to live up to Mills' predictions. Some broke down or jammed. They were never used *en masse,* and were generally deployed in out-of-the-way spots to guard bridges or narrow passes. Thus they were never really given the chance to show anything of their real potential. In August 1865 most of those that were left with the Union forces were sold off in a sale of surplus ordnance. Thirteen of them fetched a mere $500 each.

Almost all the early inventors were torn between the

desire for money and recognition and patriotic feelings, not least Richard Jordan Gatling, one of the few machine gun pioneers whose name has remained with us. There is justice in this historical assessment for the Gatling was by far the most reliable machine gun to have been produced to that date. Although it was fundamentally an amalgamation of the principles underlying the Ager and Ripley guns, the Gatling was the first weapon to have taken full advantage of the progress in machine tooling by successfully using a method of camming that ensured positive action and certainty of fire. Gatling had started work on the gun in April 1861 and the first patent was granted in November 1862. A few months earlier a working model was ready to be used in a public demonstration before thousands of people in Indianapolis. The Governor of Indiana was there and he was so impressed with the gun's performance that he wrote to F.H.Watson, the Assistant Secretary for War, to urge him that the Gatling be officially allowed to prove its worth. The *Indianapolis Evening Gazette* was even more impressed, particularly as regarded the enormous financial savings it was thought would accrue to the users of such weaponry. As they pointed out:

It takes from three to five men to work the gun to its full capacity, and it is estimated that two of the guns are fully equal to a regiment of men. One of these guns with its appendages ready for action, costs about $1,500. A regiment of men ready for the field, costs about $50,000 and it takes $150,000 to keep a regiment in service twelve months. It will be seen from the above that it would be a great economy to use the Gatling gun.[5]

All this was most gratifying to Gatling who in 1844 had moved from his birthplace, North Carolina, to Indiana in the hope that the North's greater supply of technical know-how and investment capital could help him to promote his various inventions. Gatling's faith in the unlimited benefits of technological and industrial progress was never more clearly demonstrated than in his explanation of the reasons for first inventing the gun. In a letter of 1877 he said:

It may be interesting to you to know how I came to invent the gun that bears my name . . . In 1861, during the opening events of the war . . . I witnessed almost daily the departure of troops to the front and the return of the wounded, sick and dead. The most of the latter lost their lives, not in battle, but by sickness and sickness incident to the service. It occurred to me that if I could invent a machine – a gun – that would by its rapidity of fire enable one man to do as much battle duty as a hundred, that it would to a great extent, supersede the necessity of large armies, and consequently exposure to battle and disease would be greatly diminished.[6]

But even though he moved north the better to develop and promote his weapon, Gatling could never forget that he was a Southerner, though he certainly tried to give the impression that he had. In February 1864 he wrote a letter to Lincoln:

> Sir, Pardon me the liberty I have taken in addressing you this letter. I enclose herewith a circular giving you a description of the 'Gatling Gun' of which I am inventor and patenter. The arm in question is an invention of no ordinary character. It is regarded, by all who have seen it operate, as the most effective implement of warfare invented during the war, and *it is just the thing needed to aid in crushing the present rebellion* . . . Messrs. McWhinney and Rindge – partners of mine in the manufacture of the gun – are now in Washington with a sample gun and it is hoped ere long to hear of its adoption by the War Department. Its use will, undoubtedly, be of great service to our armies now in the field. May I ask your kindly aid and assistance in getting this gun in use? I know of a truth that it will do good and effective service. Such an invention, at times like the present, seems to be providential, to be used as a means in crushing the rebellion.[7]

This wooden model of Richard Jordan Gatling's hand-cranked gun was submitted in 1862 by the inventor to the U.S. Patent Office. Gatling received U.S. Patent 36,836 on 4 November 1862 for this six-barrel design. Overall length of this dull green model is 36 inches, with 12-inch diameter wheels.

Yet at the same time that Gatling was thus praising his gun as a divine intercession in the Civil War it seems that he was in fact an active member of the Order of American Knights, a secret group of Confederate sympathisers, aiding the Southern cause by various acts of sabotage. The commanding general of the District of Indiana described Gatling as one of the most dedicated and dangerous members of the whole organisation. It is even alleged that Gatling chose to locate his actual factory in Cincinnati, only divided from the South by the River Ohio, so that if his gun ever went into quantity production it would be within easy reach of Southern raiding parties.

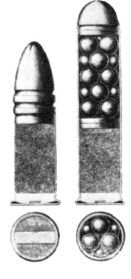

In fact, Gatling's faith in the greater business opportunities within the Northern states proved to be misplaced. Neither government nor army officials seemed to be very eager to adopt his gun. In 1863 Gatling and Rindge made many trips to Washington to argue their case but their pleas fell on deaf ears. Things became so desperate that the inventor was reduced to the most dubious of stratagems. One general at least had come out in favour of the gun. In a letter of March 1863 Major-General H.G.Wright wrote to Ripley to say that 'as a device for obtaining a heavy fire of small arms with a very few men, the Gatling seems to me admirably adapted to transport steamers plying along the western rivers, where infantry squads are needed for security, against guerrillas and other predatory bands.' Ships were in fact allowed to order Gatlings in July 1863, but few actually did so. In desperation, using a pseudonym, Gatling took a leaf out of the Major-General's book and himself wrote to the *Cairo Daily News* pointing out that 'for guerrillas on the Mississippi it will prove a very valuable weapon, and do more to drive them from attacks on our boats than any other means now used.'[8]

1 inch cartridges for Model 1865

Despite rebuffs in America Gatling's faith in his gun remained undaunted and he began to look for some foreign interest in his weapon. In October 1863 he wrote to Major Maldon of the French Artillery telling him of the terrible power of the gun and enclosing a full and accurate description of it. He further suggested that if the Major thought it ethical he might like to show the drawings to the Emperor Napoleon III. The French showed interest and Gatling immediately forwarded further particulars and a collection of endorsements from civilian and military figures who had seen the gun in action. The French had also requested a sample weapon for testing. This Gatling said he was unable to send but calmly went on to inform them that he would be more than happy to fill a minimum order of one hundred guns. The French swiftly lost interest. This was probably for the best. Gatling would doubtless have been able to adapt his conscience to the ethics of selling a hundred such weapons to a foreign power. But shortly afterwards the American Government placed an embargo upon the export of all arms and munitions.

Gatling never ceased to try and make improvements to his gun, and a new model was submitted to the Ordnance Department in 1865. By that time the Civil War was almost over, and perhaps for this reason, the inventor felt that his claims ought to be all the more extravagant. A publicity broadsheet of August of that year informed the world that:

> The gun can be discharged at the rate of *two hundred shots per minute*, and it bears the same relation to other firearms that McCormack's Reaper does to the sickle, or the sewing machine to the common needle. It will no doubt be the means of producing a great revolution in the art of warfare from the fact that a few men with it can perform the work of a regiment.[9]

In order to ensure that this gun was built to the highest standards of precision Gatling severed his partnership with McWhinney and contracted with the Cooper Firearms Manufacturing Company of Philadelphia for the production of the new model. Those pieces made in 1865 and 1866 do certainly seem to have represented a marked improvement over earlier models, and the gun finally began to make some impact upon the market. In 1866 it was adopted by the United States Army. In 1867 the British took it up, and the Japanese also bought a few models. In 1868 the Russians took delivery of twenty and contracted to buy a further hundred. Within two years, not unnaturally, the Turks came to feel that they too must have their own quota of Gatlings. In 1873 the Spanish ordered fifty Gatlings, delivery to be made in Cuba. At last the company began to prosper, for though none of these orders might seem particularly large by today's standards, Gatling was certainly not one to try and work on a low profit margin. The Russian guns, for example, cost $710 each to manufacture whilst the price paid by the Imperial Government was $1,500. Gatling himself made his business philosophy quite clear. In a letter to the Secretary of the Gatling Gun Company in 1874 he advised that: 'Our best policy will be to keep up the prices of the guns and give liberal commissions.' In October of the following year he said: 'We ought to give ten per cent commission on the guns – such a commission will make agents and gun men, consuls etc., whom we enlist in our interest work energetically in getting orders.'[10]

But few other inventors in this field achieved the same success as Gatling. Of the more than eighty patents issued in the United States for machine guns between November 1862 and June 1863 only seven were actually tested by the army or

Claxton guns with the French Army

the navy, most of the others not even getting past the drawing board. In England the story was much the same. Between 1854 and 1895 231 patents were granted for inventions relating to machine guns and automatic breech

mechanisms. Once more no more than a handful of them were ever translated into reality.[11]

But whatever one might think of the rather suspect business practices of some of these early American inventors, one basic fact shines through. However objectionable some of their actions might seem, they were, in the last analysis, underpinned by an unlimited faith in the future of human ingenuity and technology. For all these people, what was at stake was not simply the opportunity to make money but also the chance to improve the lot of mankind as a whole by the application of the most advanced techniques. Certainly they also stood to make money, but this was the prerogative of the pioneer, not the basic motivation. Thus, whilst Gatling's justification for the invention of his machine gun can to some extent be ascribed to the necessary hypocrisy of a man who stands to make money out of improving the means of killing his fellow beings, it is also significant as a classic formulation of the absolute faith men in the nineteenth century had in the beneficial effects of scientific, technological and industrial progress. This kind of attitude was nowhere more clearly demonstrated than in the exaggerations of the first and greatest confidence tricksters in the machine gun field, Myron Coloney and James McLean.

Among their supposed wares were a variety of deadly machine guns, 'Battery Guns capable of firing from six

The Montigny mitrailleuse

hundred shots a minute up to two thousand shots a minute, and sweeping an area of six miles', known by such fanciful names as the 'Annihilator', the 'Pulveriser', the 'Broom' and the 'Vixen'. To promote these guns the supposed inventors produced a 200-page booklet entitled *The Imperial Edict*. Its contents are absurd hyperbole but they effectively highlight the nature of the uncritical nineteenth-century faith in the potential of technological advance, even in the field of armaments. Of one of the 'inventors' it is said: 'Hearing of the killing and the slaughter of the brave soldiers in Europe and Asia at the will of their rulers Dr.James H.McLean . . . resolved to develop such terribly destructive weapons of war . . . as would compel all nations to keep peace towards each other.' Of his partner, Myron Coloney, one read that:

> There are . . . great inventors, who, with one master stroke of genius, wipe out all past works of a class and . . . contribute to the general prosperity of the nation . . . To this class, we think, belongs Myron Coloney, one of the inventors of the . . . McLean Peacemakers . . . Coloney engaged at once with Dr.McLean . . . to superintend and develop these terrible engines of destruction, which are intended to strike terror to the heart of every enemy . . . and which will create an enthusiasm and a sense of security in every nation on this globe . . . [12]

Maxim and Zaharoff

All the early machine guns were of the hand-crank type, where the rate of fire, though fast, depended on there being someone behind the gun continuously turning a handle. But to many inventors this was not a true machine gun. Men dreamed of a weapon that would begin to fire as soon as the trigger was depressed, and would continue to do so until the trigger was again released. The idea was finally put into practical effect by another American, Hiram Maxim, in 1884, when he gave the first demonstration of a quite new type of automatic weapon. Its essential feature was that it utilised the force of the recoil to operate the ejection, loading and firing mechanisms. Once the first round was fired the whole operation of the gun was *fully* automatic.

In a letter to *The Times* of July 1915, when the deadly effect of his weapons on the Western Front must have been readily apparent, he gave the following explanation of why he first grappled with the problems of automatic fire: 'In 1882 I was in Vienna, where I met an American Jew whom I had known in the States. He said: "Hang your chemistry and electricity! If you want to make a pile of money, invent something that will enable these Europeans to cut each others' throats with greater facility." '[13] Fired by such sentiments Maxim immediately returned to London and set up a small workshop in Hatton Garden.

For the next two years he devoted all his available time and money to the development of an automatic gun. Encouragement for his endeavours was some thing less than unanimous. At one stage he found it necessary to go to the Henry Rifled Barrel Company to obtain some reliable barrels for his prototype. The Superintendent there was either of a very gloomy disposition or, what is more likely, was very aware of the undesirability of encouraging competition in the firearms business. His advice to Maxim was emphatic:

> Don't do it. Thousands of men for many years have been working on guns; there are hundreds of failures every year; many engineers and clever men imagine that they can make a gun, but they never succeed; they are all failures, so you had better drop it, and. not spend a single penny on it. You don't stand a ghost of a chance in competition with regular gunmakers – stick to electricity.[14]

But Maxim was able to ignore these doom-laden prognostications, and soon his first model was ready for demonstration. Its efficiency and reliability were manifest from the very start and Maxim was quickly playing host to a most distinguished cavalcade of visitors. These included the Prince of Wales, the Duke of Cambridge, the Duke of Edinburgh and Sir Garnet Wolseley, one of the foremost military men of the day. The latter's first reaction to a demonstration of the gun made a fitting summary of this whole first period of machine development: 'It is really wonderful, you Yankees beat all creation. There seems to be no limit to what you are able to do.'[15]

But though Maxim managed to impress his official visitors, all his creative genius would be somewhat wasted if he could not manage to actually sell his gun. And here he came up against a major stumbling block. Though his weapon was undoubtedly the best on the market, he found himself beaten at every turn by the representative of another armaments firm, a man whose sales techniques have rarely been paralleled for their sheer unscrupulousness. This was Basil Zaharoff, an East European of obscure origins, who was at that time the European representative for the Nordenfelt Company.

The Nordenfelt machine gun had been invented by a Swede, Helge Palmcranz, and Nordenfelt himself was a banker who set up a factory in England to exploit the idea. In fact, the gun itself, a ten-barrelled, hand-cranked affair, offered no advantages over the Gatling, let alone the new Maxim. Indeed machine gun development in general in Europe at this time had been of little consequence. To a large extent this was attributable to the attitudes of the various European military establishments, and I shall be going on to deal with this in detail in the next chapter. But there were other contributory factors. The most important of these was the fact that the European machine tool industry always lagged severely behind that of the United States. Thus by the time Europeans actually had the means to produce reliable automatic weapons, the key innovations had already been made by American inventors. A major explanation of this laggardly attitude towards adopting the new methods of precision engineering was undoubtedly the traditional vested interests of the skilled European gunmakers. In Britain, for example, Henry Maudsely, in 1810, had built a set of block-making machinery for the Portsmouth dockyard which incorporated various early types of machine tools. However, 'although celebrated widely in print, and in operation for more than a century, the Portsmouth block machinery made an unaccountably slight impression upon British gunmakers and others who might have adapted the system to their manufactures.'[16] Similarly, although the British Standard Arms Company was set up as late as 1861, it still for many years did most of its work by means of skilled hand labour. Whatever the reasons, it is an indubitable fact that the European contribution to innovations in machine gun technology was minimal. In this respect it is instructive to note that, in 1914, the two major industrial powers of Europe, Britain and Germany, were both using machine guns based upon the Maxim patents, Vickers having obtained them in 1892, Krupps in 1893.

Occasionally, however, through pure salesmanship, a European was able to make some impact upon the machine gun market. Between 1877, when Zaharoff first joined the company, and 1884, he and Nordenfelt were reasonably successful in promoting their gun amongst the European powers. Of Nordenfelt it has been said: 'He proved himself one of the world's greatest salesmen, as, by sheer merchandising ability, he promoted successfully a multi-barrel weapon inferior to half a dozen other guns available at the time.'[17] But compared to Zaharoff the company's owner was a mere novice.

The salesman really began to show his mettle when it had

become apparent that the advent of the Maxim gun rendered all existing automatic weapons completely obsolete. In 1885, on hearing that the Nordenfelt had won a competition organised by the Italian Navy to find the best quick-firing gun, Maxim had immediately taken his own weapon to Spezia and proved it far superior to any of its rivals. Soon after this Maxim went to a machine gun trial in Vienna, which was to take place in the presence of Archduke William. Zaharoff realised that if such a distinguished personage ever actually saw the Maxim in operation, Nordenfelt's days in Austrian service might soon be over. So he immediately went to see the Archduke, who later related that he was told by Zaharoff that 'the weather was very hot and he advised me very strongly not to go thirty miles into the country and expose myself on the hot Steinfeld for nothing. He said the Maxim gun never works and you will be greatly disappointed.'[18] But the Archduke chose to ignore this solicitous advice and to expose his royal personage to the deadly heat. He was immediately impressed by the Maxim's capabilities and a second test was organised. This was to take place in front of the Archduke and Emperor Franz Joseph himself. Maxim was never one to miss such a grandstand opportunity. After successfully firing off several hundred rounds he concluded the demonstration by using his gun to spell out the Emperor's initials on the target. The latter was clearly impressed, but as always Zaharoff kept his head. The superiority of the Maxim could not be gainsaid so he circulated among the many journalists present and pointed out that it had been the Nordenfelt that had been responsible for the *coup de théâtre* that they had just witnessed.

A third demonstration in Vienna was arranged. This time, or so it is alleged, Zaharoff resorted to cruder methods. He came to London and bribed a milling cutter in Maxim's works to ruin a piece of the gun's casing and to rivet a replacement on in such a way that the weapon would jam as soon as it started firing. Not content with this Zaharoff also took steps at the Austrian end to sabotage any possibility of a successful trial. According to one Austrian officer who was in service at the time; 'The story is that he bribed the workers in the Maxim factory in England to send out a gun that wouldn't fire properly. Later there was a story that some young Austrian officers were mixed up in the affair . . . They were inexperienced and let themselves be gulled by this competitor into bringing unsuitable cartidges for the gun or something like that.'[19] Zaharaoff's perseverance paid off. Though it was impossible to completely conceal the merits of the Maxim, these tactics, coupled with persistent whisperings in the ears of Austrain officers and bureaucrats, kept the eventual order down to a modest 160 guns.

But it was a victory that Zaharoff must have come to rather regret. In 1888 Nordenfelt and Maxim amalgamated to found the Maxim-Nordenfelt Guns and Ammunition Company. Zaharoff was retained as the European representative and his previous campaign of sabotage and slander must have greatly hindered his efforts to reawaken interest in the Maxim gun. Certainly the new company did not do very well for the first two years of its existence. Even in 1894 it lost £21,000 and the figures for the following year, though better, still showed a deficit of £13,000. But then things began to pick up a little, and some of the credit must go to their indefatigable salesman.

Maxim and Nordenfelt had decided to merge their companies so that they would be able to compete with the great European armaments manufacturers such as Krupps, Schneider-Creusot and Armstrong in their ability to offer credit to potential customers. Zaharoff took the idea one stage further and regularly tried to bribe key personnel in the bureaucracies of various European powers. The story is told of his visit to the War Ministry in an unnamed Balkan state. The particular functionary with whom Zaharoff had to deal was not very receptive to the idea of buying any machine guns. Zaharoff was not to be put off. As he was preparing to leave he told his host that they could pursue the discussion at a later date. 'Let us,' he said, 'talk about it tomorrow, Thursday.' The potential buyer looked at him curiously. 'But tomorrow's Tuesday.' 'I can assure you,' replied Zaharoff looking him straight in the eye, 'that tomorrow is Thursday. In fact, let us bet on it. Shall we say ten thousand francs?' Being a sporting gentleman the official accepted the wager with alacrity. When he had pocketed his winnings the Balkan Government he represented became the proud possessor of a large consignment of Maxim guns. On a different occasion, in yet another nameless Balkan state, Zaharoff was once again running into difficulties with the sale of his wares. When negotiations seemed to have reached a deadlock, he offered the major with whom he was dealing a cigarette from his case. In it, along with the cigarettes, was a thousand rouble banknote. The major took it without any change of expression, and then asked if he might have another cigarette.

But, for reasons that will be explained in the next chapter, sales of machine guns never really took off until the beginning of the First World War. Up until that date, although people marvelled at the sheer efficiency of such weapons, they were never able to appreciate the extent to which they might be used on the battlefield. The Browning machine gun, which made use of the escaping muzzle gases to operate the mechanism, offers a good example of this. Browning asked the Colt Company if they were interested in developing and manufacturing the gun he had patented, in 1890. They wrote to say that they were not overly enthusiastic. The only gun they had handled to date had been the Gatling, and they pointed out that the sales drive had been expensive and relatively disappointing. But Browning has left a very picturesque description of their reaction to an actual demonstration of the prototype:

> On reaching the firing range I quit wishing that Charlie and I had changed our shirts that morning, and before anybody had time to say much, we had the gun on its mount, banging away into one of the firing tunnels . . . I ran the two hundred rounds through so fast nobody could think . . . When the last empty shell spanged on the floor, with not a hitch in two hundred, Hall and his men were too bug-eyed to see the hammer marks on the gun. They didn't look so deep to me for that matter. It was comical. You know how it is in a circus when a clown stumbles over everything and then suddenly turns into the star acrobat of the show . . . The changed expression of Hall and his men put a pound of fat on my ribs.[20]

Colt's immediately agreed to develop the gun, yet until America actually entered the First World War their vigorous sales effort made little impact on the American or European markets. Until 1914 the maufacture of machine guns merely meant steady but unspectacular business.

But slow and steady profits were not enough for certain of the arms manufacturers. There has been much speculation, particularly between the two world wars, about the role of such people on promoting the arms race that led up to the carnage of 1914–18. This is not the place to examine such a broad issue, but there does seem to be some evidence of an attempt to exacerbate the tensions between the European powers. At least one of these attempts concerned machine guns. In 1908 two members of the Reichstag, Liebknecht and Erzberger, brought that assembly's attention to a letter that had recently been sent to Paris by the Deutsche-Waffen-und-Munitions-Fabriken. Addressed to two of the firm's representatives in Paris, the letter asked that there be inserted in the paper *Figaro* an item 'which would go something like this: "The French Army Command has resolved to speed up considerably the equipment of the army with machine guns and to provide double the number originally anticipated." '[21] In fact such a news item would have been entirely bogus, for the French at that time had not decided upon any increase in their maching gun capability. For this reason the representatives were unable to prevail upon *Figaro* to run such a story, but there did appear a series of articles in that paper, *Matin* and *Echo de Paris* which made extravagant claims about the superior merits of French automatic weapons. Whereupon the Chancellor was rudely interrogated by one Deputy Schmidt, a leading member of the armaments lobby, and the Reichstag voted, without debate, a big increase in funds for machine gun development. Whereupon the French . . .

However, once the First World War had actually begun such behind-the-scenes interference became quite unnecessary. The ghastly logic of truly industrialised warfare created a demand for machine guns that was beyond the wildest dreams of any of their makers. If one bears in mind that in the ten years prior to 1914 Vickers was supplying the War Office with just under eleven machine guns per year, and that in September 1914 they were only able to supply a maximum of ten to twelve per week, the following figures should give some idea of the staggering increases in production occasioned by the First World War. Thus, in 1915 Vickers produced 2,405 guns; in 1916, 7,429; in 1917, 21,782; and in 1918, 39,473. During these same four years there were also produced for the British Army 133,196 Lewis guns and 25,379 Hotchkiss guns. All in all a grand total of just under a quarter of a million machine guns.

Such a huge output naturally brought in its wake vast profits. This description of the fortunes of the Hotchkiss company, the principal suppliers of the French Army, is fairly typical of all armaments firms during the period:

> Initial development had proved so expensive that in 1912 it was necessary to reduce the capital from six to four million francs . . . From the outbreak of the war Hotchkiss was the Allies' principal supplier of machine guns. The factory at Saint-Denis became too small, and another was set up at Lyon in September 1914. Two new ones on the outskirts of Lyon were added in the years following. Also the British Government persuaded Hotchkiss to set up a factory at Coventry . . .
> The profits occasioned by such massive mass production surpassed anything seen in any other of the armaments firms in the whole of Europe. On two successive occasions the registered capital was doubled, without anyone having to invest a penny. Such huge

reserves had been built up that the new capital could be offered free to the shareholders . . . Moreover, the dividend rose from eight to a hundred francs, in other words to one hundred per cent . . . Once the war was over, the company, having sold its factory in England, was able to pay back to the shareholders, in cash, all the capital tied up in their shares, even that which had never been paid in the first place. Thus such transactions created an appreciable number of millionaires who owed their fortunes to machine guns.[22]

The extent of the profits to be made is indicated by the fact that over the war years the Ministry of Munitions managed to reduce the price paid for a Vickers machine gun from £167 to £80, and the price of a Lewis gun, manufactured in England by BSA, from £175 to £80. BSA had in fact refused to manufacture Lewis guns at any price until the government had assured them of a massive enough order to pay the company for building an extension to their existing plant.

There were also large profits to be made in America, though here the patent holders, if not the maufactureres themselves, displayed a greater degree of patriotic magnanimity than was shown by their counterparts in Europe, or indeed their predecessors in their own country. At first the American forces were almost entirely dependent upon European weapons, and thus fell victim to a particularly brazen piece of exploitation. One of the guns in the American armoury was the Chauchat, a French development that 'was probably one of the crudest, most unreliable and cheaply made guns ever to come into service.'[23] Between December 1917 and April 1918 the Americans purchased 37,864 of these guns. This figure represented the equipment of nine divisions of infantry and was in fact double the number required for that many men. The reason for this massive over-ordering was that a good half of the guns were thrown away by the troops because they were completely useless. The American authorities could do nothing about this because under the terms of their contract with the French the weapons were to be made and inspected in France, and the Americans were obliged to accept and pay for them once they had 'passed' inspection.

But towards the end of the war the Americans managed to get their own production lines rolling and vast numbers of Lewis guns and Browning medium guns and automatic rifles were produced. Browning's royalties on all the guns produced during the war should have amounted to a little over $12,700,000 but in actual fact he declared himself satisfied with only three-quarters of a million dollars. Colonel Isaac

Lewis, the inventor of the Lewis gun, went even further and refused to accept any royalties at all. As he said when he sent back to the Secretary of War the first of many certified cheques: 'I will not accept one cent of royalty for a single Lewis gun purchased by the government of my country.'[24] All together, he returned over one million dollars to the United States Treasury. But some people at least were a little suspicious of such generosity. Colonel Crozier, the Army's Chief of Ordnance, was opposed to accepting the first cheque on the grounds that it could be construed as a bribe from the Savage Arms Company, the actual manufacturers of the gun.

These then were the pioneers of machine gun development. It might be felt that this chapter has had a little too much American bias. But the machine gun was an essentially American invention. Not simply because the four greatest names of machine gun history – Gatling, Maxim, Browning and Lewis – were Americans, but also because it was in America that were first developed the material conditions that made automatic fire a feasible proposition. On the one hand these conditions fostered the development of new tools and skills, and on the other they gave rise to a new type of warfare, the American Civil War, in which machine guns might be a necessary weapon. Finally they brought forward a brash breed of entrepreneurs who were determined that this time at least the machine gun pioneers would get their just reward, in terms of both financial gain and due recognition. For all these reasons it was in America that the machine

William J. Browning with his invention

41

gun was taken out of the footnotes of firearms history and made worthy of being displayed in the military marketplace.

Machine Guns on the Home Front

Yet the story of the connection between American industrial development and the machine gun is not quite finished. Machine guns were not only a passive product of the progress of industrial capitalism, they were also actively deployed in its defence. In America, in the late nineteenth and early twentieth centuries, the machine gun was a standard weapon in the bitter struggle between management and organised labour.

As early as 1863 H.J.Raymond, the owner of the *New York Times*, had bought three Gatling guns to protect his offices against feared attacks by mobs of people protesting against the Conscription Act of March of that year, of which the *Times* had come out in support. Luckily for all concerned the Gatlings were never needed. In Pittsburgh, in 1877, during the Great Strike, the Philadelphia National Guard was called out and its equipment included at least one Gatling. It too was never used because the Guardsmen broke and fled rather than have to fire upon their fellow citizens. In 1891 the miners in Briceville, Tennessee, went on strike against the use of cheap convict labour, and a fortified post was set up there by the Governor, manned by 175 Guardsmen and a Gatling. At the siege of Carnegie's Homestead steel mills in 1892, 8,000 Pennsylvania National Guardsmen with several Gatlings were deployed against the strikers. The following quotation from the *Chicago Times* of 1896 shows the extent to which machine guns had already become a standard weapon in the war against labour:

> Within the past forty-eight hours nearly $2,000 has been subscribed by various members of the Commercial Club, with which it is promised to purchase some sort of machine gun for the First Infantry, Illinois National Guard. The idea was suggested at inspection and drill of the regiment Tuesday night and was readily adopted when it was hinted that in case of a riot such a piece would prove a valuable weapon in the hands of the Guardsmen.[25]

Machine guns continued to figure largely in industrial disputes in America right up until the eve of the Second World War. They were used in West Virginia by National Guardsmen in the strikes of 1920 and 1922. During the longshoremen's strike in San Francisco soon afterwards the Governor of California sent in 1,700 National Guardsmen who set up machine-gun nests all round the city. As the leader of the unofficial union that was heading the strike said, on its collapse: 'We cannot stand up against police, machine guns and National Guard bayonets.' In the 1930s the Thompson sub-machine gun became the favoured weapon. General John T.Thompson, the gun's inventor, whose business was going through a very bad time, worked assiduously to interest big business in the merits of his weapon. Working through Federal Laboratories, a so-called 'protection engineering' firm, Thompson managed to dispose of several hundred guns.

But it was in Colorado, in the first two decades of this century, that machine guns really came into their own as an alternative to collective bargaining. The economy of the state was dominated by the Colorado Fuel and Iron Company whose miners were forced to work under the most appalling conditions. They lived in tents or broken-down cabins, worked in totally unsafe mines, were paid in company scrip, and were surrounded by armed guards in case they wished to protest. In Cripple Creek, in 1903, the miners had gone on strike. They were quickly rounded up and imprisoned, without trial, in large bull pens. Eventually a Colorado judge ordered that habeas corpus proceedings be instituted on behalf of these prisoners. But he reckoned without the local National Guard. Its commander was a certain General John Chase, all of whose elder relatives had fought in the American Civil War, and who felt called upon to emulate their example. He had been denied the opportunity of fighting in the Spanish-American War because of some irregularity in his commission as major, so he decided to test his martial abilities against the local workers. Urged on by the mine-owners he surrounded the Court House at Cripple Creek with National Guardsmen, posted riflemen on the roofs roundabout, and set up a Gatling gun in the street outside. The legal proceedings were broken up and the prisoners taken back to the bull pen.

In 1912 trouble broke out again at a place called Cabin Creek. This time private company guards were used to break up the strike, but they too had a Gatling gun at their disposal. It was never used but did provide the opportunity for some courageous radical rhetoric. Mother Jones, one of the miners' most stalwart leaders, arrived in Cabin Creek in her buggy, drove straight up to the armed guards, and placed her hand on the muzzle of the Gatling. As she herself described it in a later speech: 'There was a gang of those guards with a Gatling gun . . . and they said, "Take your hand off that gun", and I said, "Oh, no, sir, my men made that gun, sir." '

The confrontation between miners and guards lasted right through 1912 and 1913, and over the months the mining companies stockpiled more and more machine guns. One of the most prominent suppliers of these weapons was the Baldwin-Felts Detective Agency under the local leadership of Albert and Dee Felts. The former, 'functioning as a kind of military overseer for the mine-owners, arranged the importation of at least eight machine guns from West Virginia, where some of them had recently seen service against striking miners.'[26] Nor were they kept merely for show:

> Firing a hailstorm of bullets into the little town became a hobby with the gunmen. There were several vantage points from which the machine guns could be trained on the town, and frequently showers of leaden hail would sprinkle the tent colony and surrounding buildings . . . After an attack on gunmen at Mucklow machine guns opened up from below and rat, tat, tat a whine of bullets answered the volley of the strikers . . . One machine gun was among some piles of railroad ties in the valley, another was in an empty box-car on the siding. Several of the best shots were designated to silence the machine guns, if possible.[27]

Later the mine-owners got hold of an armoured train, equipped with machine guns and known as the 'Bull Moose Special', which patrolled and regularly strafed the area around Cabin Creek and Paint Creek. One of the strikers, Ralph Chaplin, composed a song whose last verse adequately sums up the feelings of the desperate miners:

> They riddled us with volley after volley;
> We heard their speeding bullets zip and ring,
> But soon we'll make them suffer for their folly –
> Oh, Buddy, how I'm longing for the Spring.

But the worst incident of all took place at Ludlow, in 1913. There the mine-owners had another armoured train equipped with machine guns, known this time as the 'Death Special'. This was manned by private gunmen but it was the National Guard who were responsible for the ultimate outrage. For no reason they one day surrounded the Ludlow tent colony with machine guns and leisurely began to rake the whole area with bullets. After a few minutes the intensity of the firing caused the tents to catch fire and almost the whole area was razed to the ground. The combination of bullets and flames accounted for 36 people dead and over 100 injured.

Notes

1. D.J.Boorstin, *The Americans*, Pelican Books, Harmondsworth, 1969, vol.2, p.47.
2. Ibid., p.51.
3. Ibid., pp.48–9.
4. W.Millis, *Arms and Men*, New American Library, New York, 1963, pp.102–3.
5. P.Wahl and D.R.Toppel, *The Gatling Gun*, Herbert Jenkins, London, 1966, p.18.
6. Ibid., p.12.
7. G.M.Chinn, *The Machine Gun*, Bureau of Ordnance, Department of the Navy, US Government Printing Office, Washington, 1951, vol.1, pp.52–3.
8. Wahl and Toppel, op.cit., p.19.
9. Ibid., p.21.
10. Ibid., p.70.
11. See Chinn, op.cit., p.117, and W.R.Lake, *Machine Guns and Automatic Breech Mechanisms*, Haseltine Lake and Co., London, 1896, pp.9–19.
12. Chinn, op. cit., pp.104–6.
13. Lt.-Col.G.S.Hutchison, *Machine Guns: their History and Tactical Employment*, Macmillan, London, 1938, pp.48–9.
14. H.Maxim, *My Life*, Methuen, London, 1915, p.163.
15. Ibid., p.164.
16. M.Kranzberg and C.W.Pursell (eds.), *Technology in Western Civilisation*, Oxford University Press, New York, 1967, p.278.
17. Chinn, op.cit., p.110.
18. R.Neumann, *Zaharoff the Armaments King*, Allen and Unwin, London, 1938, p.89.
19. Ibid., p.90.
20. J.Browning and C.Gentry, *John M.Browning: American Gunmaker*, Doubleday, New York, 1964, p.152.
21. D.McCormick, *Pedlar of Death*, Macdonald, London, 1965, p.84.
22. R.Lewinsohn, *Les profits de guerre à travers les siecles*, Payot, Paris, 1935, pp.146–7.
23. F.W.A.Hobart, *Pictorial History of the Machine Gun*, Ian Allen, Shepperton, 1971, p.95.
24. Chinn, op.cit., p.297.
25. J.Brecher, *Strike*, Straight Arrow Books, New York, 1972, p.43.
26. G.McGovern and L.Gutteridge, *The Great Coalfield War*, Houghton Mifflin Co., Boston, 1972, pp.117–8.
27. F.Mooney (ed.J.Hess), *Struggle in the Coalfields*, West Virginia University Library, Morgantown, 1967, p.34.

III *Officers and Gentlemen*

'Our cavalry must be officered. We may require from the candidates either money or brains. The supply is most unlikely to meet the demand if we endeavour to exact both.'

Mr. Akers-Douglas, 1903.

So far I have only dealt with the actual appearance of the machine gun on the market and have tried to explain why it should have appeared at the time it did. A part of this explanation involved the American Civil War and the extent to which it can be regarded as the first example of an industrialised conflict, in which technological advances dictated much of the actual conduct of the war. In this kind of mass warfare weapons of mass warfare became suddenly relevant, and it was no coincidence that machine guns should first appear on the battlefields of the war between the states.

The reason the Civil War so quickly absorbed the productive capacities of both sides was that it was a bitter struggle to the end, in which there could only be a victor and a vanquished. But another reason why it had assumed a much more total nature than, say, the Crimean War, less than ten years before, or Bismarck's wars that followed it, in 1866 and 1870–71, was that the American standing army was very

Hiram Maxim letting the Prince of Wales have his turn

47

small and there were few men with any rigid conceptions of what war *ought* to be like. Both soldiers and civilians alike lived in something of a military vacuum and were much more ready to adapt all kinds of available techniques for the furtherance of the war effort. The American Civil War saw the first appearance of a whole host of military items, such as the machine gun, rifled weapons, breech-loading and magazine arms, the land mine, the field telegraph, steam-driven ironclads, and even a crude type of submarine. It also saw the first widespread use of railways to move troops around, and of the techniques of mass production to supply them with food, uniforms and equipment.

But this open-minded attitude to the nature of warfare did not exist to anything like the same extent in the countries of Europe. Each of them had a large officer corps whose origins went back hundreds of years, as did most of their tactical doctrines. Throughout the eighteenth and nineteenth centuries these officer corps remained effectively cut off from outside social and technical developments. Though this endowed them with a cohesive *esprit de corps*, it also made them very unreceptive to what was going on in the world outside. In Britain, for example, the officer corps was dominated by the aristocracy and the gentry. In 1875, 18 per cent of the officers were from the aristocracy and 32 per cent from the gentry. In 1912, the figures were 9 per cent and 32 per cent respectively. Even more significantly, if one places all the officers of the rank of Major-General and above in their appropriate status group one finds that in 1912, 24 per cent of them were from the aristocracy and a further 40 per cent from the gentry. In certain of the most influential units, the Guards for example, these figures were far higher.[1]

In France the army had been effectively purged of this aristocratic element during the Revolution. Yet in the following century the situation was slowly but effectively reversed, under the Bourbons, the July Monarchy and the Second Empire. Nor did the advent of the Third Republic improve matters. Ironically, as the more conservative elements were gradually pushed out of economic and political life, they turned to the army as their only haven. In 1873, for example, of the 365 officer cadets at St Cyr, 102 of them had a *de* before their surname.[2]

Similar trends were evident in Germany. Between 1806 and 1816 a concerted effort had been made by certain Prussian reformers to throw the officer corps open to all classes. Between 1857 and 1860 the reactionaries, led by General von Manteuffel, purged the army of as many non-noble officers as possible. In 1860 the Prussian officer corps was 65 per cent aristocratic. By 1913 this figure, due to the demands of a vast numerical expansion of the army as a whole, had dropped to 30 per cent, but the effect of the earlier policy was shown by the fact that 53 per cent of all officers above the rank of colonel were of noble origin. In 1902 the Kaiser made the connection between aristocracy and officer corps very clear when he supported augmented grain tariffs which would protect the economic interests of the Junkers. For, as he said, 'they are bound up with the preservation of the basis for the further existence of the officer corps.'[3]

The Contempt for Technology

It seems certain that this exclusive nature of the European officer corps had a marked effect on their attitudes to technological advance. Vagts has made the point well with reference to the first half of the nineteenth century:

> (There was) a decided hardening of the arteries . . . in the general staffs, evidenced in their reluctance to consider technological innovations; how utterly averse they were to changes in material was discovered to their great chagrin by industrialists like Colt, Krupp and Whitworth, and the military inventor and officer Werner Siemens grew so disgusted with the Prussian service that he gave it up to go into industry . . . Especially in the first half of the nineteenth century . . . the officer remained a romantic in the industrial age.[4]

General Manteuffel gave a very clear example of this contempt for all things technical when he attacked the whole notion of schools 'where often enough haughty teachers, as a rule full of hostility to war and the better classes, proud of their fancied scholarship, in brutal ways kill the feeling of honour and, saturated with the destructive tendencies of the time, almost exclusively rationalist, educate for everything else but character.'[5]

But there was more to the officers' attitude than a mere inability to comprehend technological progress. Because of their aristocratic origins many of the officers were, as Vagts put it, romantics in an industrial age. Their social isolation limited them to a conception of war as it had existed in a previous century. They still believed in the glorious cavalry charge and, above all, the supremacy of man as opposed to mere machines. Certainly they acknowledged that soldiers got killed by firearms, but they were never prepared to admit that advances in technology had reached such a level that the staunchest assault by the best of troops could be brought to nothing by modern weapons.

Certain thinkers did see that this was so. Engels, for example, stated flatly that:

> Force is no mere act of will but calls for very real conditions before it starts to work, in particular tools, of which the perfect one overcomes the imperfect ones; that furthermore these tools must be produced, which means at the same time that the producer of more perfect tools, *vulgo* arms, beats the producer of more imperfect ones.[6]

But the officers of the nineteenth century completely missed this point about the increasing dominance of the tools of war. For them war still *was* an act of will. Military memories and traditions had been formed in a pre-industrial age when the final bayonet or cavalry charge might be decisive. For them, in the last analysis, *man* was the master of the battlefield. In the pages that follow there will be numerous examples of this faith in outmoded tactical conceptions based upon unreal attitudes to the role of man alone on the battlefield. At the moment I wish simply to indicate the existence of the basic contradiction between theory and reality. Industrialisation and technological progress had in fact made man secondary to machines and had robbed him of his primacy on the battlefield. But in the military hierarchies of Europe the traditional ideas lived on, and the whole period up to 1918 and even beyond is dominated by this basic contradiction. In a stimulating study of the fictional treatment of war at this time, I.F.Clarke is drawn to just this conclusion:

> The great paradox running through the whole of this production of imaginary wars between 1871 and 1914 was the total failure of army and navy writers to guess what would happen when the major industrial nations decided to fight it out. Even when one takes account of the fact that many of these writers were presenting a special case for changes in equipment or organisation by projecting success or disaster into the future, it still remains true that the intense conservatism of the armed forces and years of studying pre-technological battles from Cannae to Waterloo had induced a habit of expecting that wars would continue to be more or less as they had always been. In consequence the naval and military prophets generally saw war as an affair of adaptation and improvisation. They rarely thought of what their new equipment might do. None of them ever

seems to have imagined that technology might be able to create new instruments of war.[7]

Clarke goes on to underline the point about the European officer corps reflecting, even in the nineteenth century, the old aspirations and ideals of a defunct social class: 'In their own strange way these writers were trying to create a Beowulf myth for an industrial civilisation of ironclads and highspeed turbines, a new and violent *chanson de geste* for an age of imperialism, told in the inflammatory language of the mass press . . . In the closing years of the nineteenth century the aggressive nation states of Europe had everything on their side except common sense.'[8]

All this boded ill for any hopes of the military establishment's taking kindly to the introduction of the machine gun. For they were both ignorant and suspicious of all the great advances in firearms manufacture that were made in the nineteenth century. Over the years these developments – the machine gun, breech-loading rifles since the 1860s, magazine rifles from the 1880s, the use of smokeless powder to facilitate aiming from 1885, quick-firing artillery developed in the 1890s – all these had gradually made it impossible for troops to cross open ground in any number.[9]

This point had become apparent even in the American Civil War when the troops had much recourse to entrenchments to protect themselves from enemy fire. But most Europeans, even those who actually witnessed the war, did not draw the appropriate lessons. They chose to regard it as a military aberration, not likely to be repeated on the more civilised battlefields of Europe. Some lip-service was paid to the conclusions drawn by the more prescient thinkers. On hearing a lecture given by the Professor of Military History at the Staff College in England, no less a distinguished person than the Duke of Cambridge was forced to admit that 'in this war . . . and in all future wars, the spade must form a great element in campaigns . . . Now that I think is a very great and new element in the features of war, and probably the necessity for it has been greatly increased by the improvements in arms, both artillery and musketry.'[10] A few lone thinkers tried to hammer the lesson home. In Britain, in 1873, Major-General Sir Patrick MacDougall pointed out that 'a front line held by good troops undemoralised is practically unassailable under the present conditions of fire.'[11] In France, F.Lecomte, in a book about the Civil War, made much of the fact that repeating rifles, large calibre artillery, armoured ships and Gatling guns were the new weapons of the Industrial Revolution and as such were bound to fundamentally alter the most basic concepts of war.

In Prussia, Heros von Borcke 'took pains to examine many corpses and found so few stabbing wounds that he concluded that "bayonet fights rarely if ever occur, and exist only in the imagination".'[12] After spending twelve years examining the nature of modern warfare the Polish banker, Ivan Bloch, came to the conclusion that:

> War will become a kind of stalemate . . . Everybody will be entrenched. It will be a great war of entrenchments. The spade will be as indispensable to the soldier as his rifle . . . It will of necessity partake of the character of siege operations . . . There will be increased slaughter . . . on so terrible a scale as to render it impossible to get troops to push the battle to a decisive issue. They will try to, thinking that they are fighting under the old conditions, and they will learn such a lesson that they will abandon the attempt for ever.[13]

But such warnings fell on deaf ears. The average officer was simply not receptive to any ideas that might force him to change his comfortable habits. In 1872, Major-General Sir John Maurice bemoaned the fact that 'the colonels of the regiments as a rule are fighting hard to stand to the old ways. That queer animal in the matter of books, the British officer . . . as a rule does not know where to get books . . . He hates literary work, even in the form of writing letters, has hardly the energy to undertake it.'[14] The noted military writer, J.F.C.Fuller, gave a particularly striking example of this aversion to reading: 'In 1913, I remember a major recommending Henderson's *Stonewall Jackson* to a brother officer, and then, a few minutes later, when this book was being discussed, committing the error of supposing that the Battle of "Cross Keys" was a public house in Odiham and Jackson the name of the man who ran it.'[15]

Such unimaginative men inevitably produced unimaginative military doctrines. For them the prime task of soldiers was still to charge across open ground to come to grips with the enemy. In his *History of the American War*, Lieutenant-Colonel Fletcher of the Scots Guards completely underestimated the importance of modern firepower and his conclusions about its significance are typical of European military thought right up until the eve of the First World War. He attempted to explain the character of the Civil War in terms of the American temperament and said: 'The rapid, well-sustained attack, which in many of the great European combats has led to immortal successes, does not appear adapted to the qualities of the Federal soldiery.'[16] Even such a thoughtful student of war as Spencer Wilkinson, founder of the Manchester Tactical Society and one of the great proponents of the Volunteer system, was able to say, in 1891:

> It is true that within certain narrow limits, which can be precisely specified, the defender is strengthened by modern improvements in firearms. But it is not true that this results in a great or sudden change in the relations of attack and defence, either in regard to battle as a whole, or in regard to the general course of a campaign. There has been no revolution in tactics or in strategy, but certain modifications long since realised have become more pronounced. The balance of advantage remains where it was.[17]

It is indeed true that there was no revolution in tactics at this time. One is, however, entitled to think that there ought to have been. Colonel Richard Meinertzhagen, a regular officer in the British Army in the early part of this century has described the kind of drill that the army insisted upon until the bitter end. In 1899 he wrote in his diary: 'Today we were taught how to assault an enemy position. The battalion moved forward in tight little bunches of about twenty men each, marvellous targets for modern riflemen and machine gunners, but the drill was splendid, shoulder to shoulder and perfect line. We should have been annihilated long before we reached the assaulting line.'[18]

French tactical doctrines were even more extreme. Under the influence of people like Foch, Grandmaison and Langlois, 'who established a school of thought rivalled only by the Dervishes in the Sudan'[19], French infantry tactics came to stress the offensive to the exclusion of everything else. The French Army Regulations of 1884, for example, had demanded *closer* infantry formations than those of 1875. Those of 1895 defined future battle in these terms:

> As soon as the battalion has arrived within 400 metres of the enemy, bayonets are fixed and indirect fire . . . of the greatest intensity delivered . . . At a distance of 150 metres magazine fire is commenced, and all available reserves close up for the assault. At a signal from the colonel the drums beat, the bugles sound the advance and the entire charges forward with cries of 'en avant, a la baionnette!'[20]

Even in 1913 Joffre described the task of the French Army in these terms: 'The French Army, returning to its traditions, no longer knows any other law than the offensive . . . All attacks are to be pushed to the extreme with the firm resolution to charge the enemy with the bayonet, in order to destroy him . . . This result can only be obtained at the price of bloody sacrifices. Any other conception ought to be rejected as contrary to the very nature of war.'[21] Attitudes in Germany were similar. The Infantry Regulations of 1899, which remained substantially unchanged until 1914, had this to say about the attack:

> When the decision to assault originates from the commanders in the rear, notice thereof is given by sounding the signal 'fix bayonets' . . . As soon as the leading line is to form for the assault, all the trumpeters sound the signal 'forward, double time', all the drummers beat their drums, and all parts of the force throw themselves with the greatest determination upon the enemy. It should be a point of honour with skirmishers not to allow the supports to overtake them earlier than the moment of penetrating the enemy's position. When immediately in front of the enemy, the men should charge with bayonet and, with a cheer, penetrate the position.[22]

But the most bizarre reflection of the contradiction between the old modes of thought and the new weapons was the grim determination with which military establishments clung on to their cavalry regiments. Once again the American Civil War had shown the way must develop if it was to even try and be seriously considered in modern warfare. In a letter of 1862 Captain Edward Hewitt of the Royal Engineers regretfully observed that: 'The cavalry on both sides are merely mounted infantry. They are not taught to use the sword at all . . . They never charge or get well amongst the infantry . . .' Major-General J.H.Wilson showed somewhat more resolution in drawing the correct conclusions about the use of cavalry as dismounted infantry: 'Not until the closing days of the war did we wake up to what our experience . . . ought to have taught us . . . There are only two arms that cavalry should use in modern warfare – the repeating magazine gun . . . and the revolver.'[23]

Certainly, from 1862 onwards, there were some enlightened men who deprecated the value of cavalry upon the actual battlefield. Another military man impressed by the example of the Civil War was Major Henry Havelock. In a book called *Three Main Military Questions of the Day* he bemoaned contemporary cavalry tactics which made it 'the jangling, brilliant, costly, but almost helpless reality it is.'[24] In the wake of the Franco-Prussian War, a French officer, Colonel T.Bonie, pointed out that of the 65,000 German casualties cited in the lists of the medical corps, the sabre had accounted for a mere six dead and 212 wounded. In 1891 the debate even reached out into civilian life when *Punch* carried a cartoon portraying a discussion between an infantry and a cavalry officer.

In certain quarters there was also apprehension about the quality of the officers attracted to such an obsolete arm. In 1898 Colonel Meinertzhagen remarked that: 'The only advantage in cavalry is the smarter uniform. The two cavalry regiments that I have seen are not so efficient at their work as the infantry, and the class of officer in the cavalry does not seem to take his profession too seriously.'[25] The most damning comment of all came from Mr Akers-Douglas at the Interdepartmental Commission of Enquiry, of 1902 to 1903, on Military Education: 'Our cavalry must be officered. We may require from the candidates either money or brains; the supply is most unlikely to meet the demand if we endeavour to exact both.'[26]

But neither statistics nor bitter humour could make much impression on those with the power to make actual decisions. Certainly there was some debate but the issue was never really in doubt. Traditionalists ruled the military roost and the very essence of their traditionalism was the belief in the old aristocratic ideal of the glorious cavalry charge. This became even more true in the years just before the First World War, for, amazing though it might seem, though the War Office had been dominated by the Artillery up until about 1890, cavalry officers became dominant after this date. This became evident in the discussion that followed the Russo-Japanese War. Though, as will be seen, modern rifles and machine guns dominated the battlefields of Manchuria, 'the virtually inevitable effects of these weapons on the vulnerable target of the horse . . . were evident only to the minority with truly open minds. In postwar discussions the British cavalry concerned itself with comparatively minor issues, such as the poor horsemanship of both sides, the opportunities lost by the Japanese, for lack of cavalry, and the uncritical assertion that training in mounted infantry tactics "kills the cavalry spirit".'[27]

Thus the British Cavalry Training Manual of 1907: 'It must be accepted as a principle that the rifle, effective as it is, cannot replace the effect produced by the speed of the horse, the magnetism of the charge, and the terror of cold steel.' Luckily for the Germans, in the First World War, they used machine guns, pill boxes and barbed wire that seem to have been immune to such awesome tactics. That it took the British generals so long to get this through their heads is partly explained by the fact that nearly all of them were cavalry men. Thus Haig, in 1904, attacked a writer who 'sneers at the effect produced by sword and lance in modern war; surely he forgets that it is not the weapon carried but the moral factor of an apparently irresistible force, coming on at highest speed in spite of rifle fire, which affects the nerves and aim of the . . . rifleman.'[29] But rare were the cavalry men who came on in spite of sustained machine gun fire. Haig, above all people, should have learnt this simple lesson. Yet in 1926, in a review of a book by Liddell-Hart, Haig asserted that though there were some blasphemous spirits who thought that the horse might become extinct, at least on the battlefield, 'I believe that the value of the horse and the opportunity for the horse in the future are likely to be as great as ever . . . Aeroplanes and tanks are only accessories to the man and the horse, and I feel sure that as time goes on you will find just as much use for the horse – the well-bred horse – as you have ever done in the past.'[29]

By this time such faith in the potential of cavalry had become a little eccentric. But prior to the First World War such faith was part and parcel of the orthodox system of military beliefs. Europe as a whole was a little bewildered by the sheer speed of technological progress. And within each nation the army above all, nourished as it had been on the old ideals of personal combat and honourable death, found it most difficult to face up to a concept of war in which death struck whole regiments at a time, delivered by an enemy one could not even see. Even some of the most perceptive military critics were not immune to these fond recollections of war as it had once been. Lieutenant-Colonel Charles à Court Repington, whilst he admitted at one stage that modern science seemed to have 'outstripped the capacity of certain nations to make intelligent use of the new weapons', nevertheless spoke of 'the true cavalry spirit which scorns mathematical calculations'.[30] In the last analysis one cannot simply disregard such attitudes as mere reactionary blimpishness. In this respect one is forced to agree with Brian Bond who notes that: 'Although the motives were perhaps subconcious, such phenomena as the cult of the horse and the *arme blanche* may now be seen as a last desperate effort to withstand the depersonalisation of war.'[31]

Reactions to the Machine Gun

It is thus hardly surprising that the general military reaction to the machine gun was something less than enthusiastic. There was some support from seemingly important quarters, but in fact such proponents of automatic weapons tended to be isolated individuals or to come from groups unable to make much impact on the orthodox military establishment.

Thus, in 1879, Lord Chelmsford remarked: 'At Ulundi we ... had two Gatlings ... When in work they proved a very valuable addition to the strength of our defence ... As ... an infantry weapon ... they might, I feel sure, be used most effectively not only in defence, but in covering the last stage of an infantry attack.'[32] In 1882, speaking at the Cutlers' Company, Lord Charles Beresford asserted that: 'In my opinion, machine guns, if properly worked, could decide the fate of a campaign, and would be equally useful ashore or afloat.'[33] In 1885, Sir Garnet Wolseley felt convinced that, 'the British army has now most certainly arrived at the conclusion that we must have machine guns ... I feel convinced that the fire of this small arm, fired from a fixed carriage ... will be most effective.' On another occasion he was even more emphatic: 'The machine gun is still in its infancy. Its power when in its prime will in my opinion astonish the world.'[34] But Wolseley's optimism about the perspicacity of the British army was a little premature. For he and the other speakers cited here were outside the mainstream of military thought. Beresford was actually a Navy man, and he, Chelmsford and Wolseley had all gained their reputations in the many colonial wars that Britain waged in the nineteenth century. Though this made them popular figures with the general public it in fact reduced their influence within the military hierarchy. For, in the last analysis, the majority of soldiers regarded such imperialist expeditions as mere sideshows, and concentrated their thoughts upon the possibility of a future war in Europe. Thus, though Wolseley's name was literally proverbial – to say that something was 'all Sir Garnet' meant that it was in perfect working order – the army as a whole was somewhat contemptuous of his theoretical pronouncements:

> Though Wolseley had a great knowledge of war derived from fighting in almost every clime and under every condition, his foes – in the Horse Guards especially – denied him the right to pose as an expert on military affairs. Small wars against uncivilised nations were an irrelevance; his experience under outlandish circumstances could never be applied to an army drilled to fight a European war.[35]

So the machine gun became associated with colonial expeditions and the slaughter of natives, and was thus by definition regarded as being totally inappropriate to the conditions of regular European warfare.

Outside the circle of colonial campaigners recognition of the machine gun's potential was very limited. Only a few isolated individuals even had the slightest inkling of what such weapons might do, and the bulk of even this support was only evident in England after machine guns had been in existence for some twenty years. In the rest of Europe yet another twenty years had to elapse before the adherents of automatic firepower could find the opportunity to express their views.

One group whose support might have been thought to be of great weight in the debate about firepower was the Hythe School of Musketry. In 1897 they strongly attacked the army's neglect of the machine gun:

> Machine gun practice is not satisfactorily conducted, and this can hardly be looked for until each regiment and battalion has been supplied with the gun. The present system works badly, and the officers NCOs and men of the gun detachments very soon become rusty and forget all they have been taught on account of months – often years – passing without ever handling one of these intricate weapons.[36]

In 1907 the officer in charge at Hythe, Lieutenant-Colonel N.R.McMahon, made the most strenuous efforts to enlighten his superiors about the importance of this issue. In one lecture he said: 'Machine guns will be used in the near future in very large numbers. There need be no fear of overstating the value of these weapons. All tendencies in modern tactics . . . bring their good qualities more and more into relief.' But the authorities ignored these appeals and so McMahon, desperately aware of the crucial importance of firepower, was forced to concentrate upon trying to make the British regular soldier the best rifleman in the world. In this he succeeded, but he never regarded this as more than an unsatisfactory compromise: 'If my advice had been heeded by the Army Council, we would certainly have obtained superiority of fire with machine guns. This advice was disregarded, so we had to fall back on the rifle . . . The folly of neglecting such weapons is apparent.'[37]

If the advice of the very people in charge of light arms training was ignored it is hardly surprising that other individual supporters were mere voices crying in the wilderness. In 1882 a Captain Fosberry pointed out that:

> A most ably conducted and exhaustive set of trials has already determined that we have a system of machine guns by which one or two men can do as much

destruction as, say, forty ordinary soldiers ... And yet, though it is certain that an enemy will always do the unexpected thing, and if a European one, use machine guns against us, we neglect to acquire them for the army because the exact tactical place for the weapon is as yet undiscovered.[38]

In 1910 Allenby was still trying to hammer the lesson home: 'The question of machine guns might be studied by the cavalry nowadays, because I do not think we make sufficient use of them. The weapon is not properly understood, and I think that, whether in fire tactics or in the tactical use of the weapon, we have hardly yet made a beginning. Personally I believe it is going to have an enormous future before it.'[39] Yet on the very eve of the First World War such men had not been able to make even the slightest dent in the smug complacency of the High Command. In 1913 Meinertzhagen was still being forced to preach the old lessons:

What I should have liked to see is more automatic fire power in the hands of both the battalion commander and the company commanders ... What we want is four machine guns ... at the disposal of battalion commanders and two with each company. It would double the fighting strength of a battalion and would

(left to right) Gardner Gun, a Maxim and a Nordenfeldt

reduce its strength and expense, provided one section of each company was a machine gun section. The power of machine guns has not yet been appreciated in our army.[40]

In January 1914 J.F.C.Fuller was severely criticised for writing a paper, at the Staff College, 'whose main contention was that tactics are based on weapon-power and not on the experience of military history, and that since in 1914 the quick-firing field gun and the machine gun were the two most recent weapons, our tactics should be based on them.'[41]

In Britain not even the support of royalty was able to further the machine gun cause. The Prince of Wales had been one of the very first visitors to Maxim's factory in Hatton Garden, and he had been immediately impressed by the capabilities of the new gun. But, ironically, the only people on whom his enthusiasm seems to have had any impact were the members of the German royal family. In 1887 Prince Wilhelm, during a visit to London to attend the Golden Jubilee celebrations, had been taken to see the

An attempt to amalgamate machine guns and cavalry Danish hussar 1901

60

machine gun equipment of the Tenth Royal Hussars. He was so impressed by what he saw that he had a gun sent back to Potsdam with a British gunner assigned to teach the German troops. A few years later, as Maxim relates: 'The Prince of Wales visited the Kaiser, and when the conversation turned on arms the Prince asked the Kaiser if he had seen the Maxim gun. He said he had not, but that he had heard a lot about it. The Prince told him it was really a wonderful gun and . . . suggested that they should go out and see it.' The Kaiser was duly impressed. After the demonstration he said quite simply 'This is the gun. There is no other.'[42] After the Imperial Army manoeuvres at Metz in 1902 he is reputed to have claimed that one brigade with plenty of machine guns could undoubtedly hold an army corps in check.

Their Imperial Highnesses wielded much more influence at home than did British royalty. In the years leading up to the First World War the machine gun strength of the German Army was brought up to a much higher level than in any other army in Europe. In 1899 a four-gun Maxim battery was allotted to each Jäger battalion, and in 1902 each cavalry brigade on frontier duty received a six-gun battery. In the following year all batteries were given six guns, and between 1905 and 1908 each infantry regiment was given its own complement of Maxims. In the latter year a further fourteen million marks were set aside for machine gun experiments. Inter-regimental machine gun firing competitions were also instituted, the winner to receive a watch, given in the name of the Kaiser and inscribed with his name. The importance the Germans attached to these weapons is clearly revealed by the following extract from that section of the Field Regulations dealing with machine guns and their crews:

> The essential point for the machine gun detachment is to be able to shoot well and open fire at the opportune moment from the most favourable position, and against the most useful objective. To secure all this it is necessary to have a thorough knowledge of the gun, a very mobile detachment, leaders who possess tactical insight, and men full of initiative who, devoted heart and soul to the Emperor and the Fatherland, will exert themselves to obtain victory, even when they have lost all their leaders.[43]

Caught up in the European arms race, the French too began to display a belated enthusiasm for machine guns, and certain writers put similar stress upon the key tactical role of such weapons. In 1906 one Commandant Guérin insisted that the officer in command of a machine gun battery 'must be a man of cool character with a good eye for ground, and ready decision in choosing the fire positions and targets against which to use his gun, and in selecting the precise moment when he must change position and push forward as ground is gained.' In 1910 Commandant Lavau became quite mystical about the qualities of a good machine gunner:

> As there is a cavalry spirit and an artillery spirit which moulds the ideas and directs the actions of the cavalier and the gunner, so . . . there should be a 'machine gun spirit' in those who handle the new weapon – a spirit of enterprise and dash, of readiness to seize the fleeting opportunities that offer themselves amid the confusion of the battlefield for the sudden intervention of the death-dealing weapon.[44]

Shortly before this a certain Charles Humbert had written a book about the German superiority in machine guns called *Sommes-nous défendus?* Its effect had been such that it precipitated the threat of a vote of 'no confidence' in the Assembly. Panic-stricken, the Government immediately promised to manufacture thousands of new guns without delay. By the time of the manoeuvres of the following year each infantry regiment had its own battery of six Hotchkiss machine guns.

But in Britain, right up until 1914, the story was one of unremitting hostility. Through three major wars in which machine guns were employed, and innumerable small expeditions against helpless tribesmen in Africa and Asia, the British military authorities remained stubbornly unaware of the inevitable consequences of the perfection of automatic weapons. Just as they had been deaf to the entreaties of the handful of more enlightened thinkers, so did they close their eyes to the very evidence of the wars themselves.

Machine Guns in Action

Reactions were hostile from the very beginning. Writing of the Gatling in 1862, three British officers sent to America to report back on the course and nature of the Civil War were far from enthusiastic:

> We saw some practice at 250 yards range against a target, with this gun, which was very bad; this appeared to be the fault of the ammunition, as the bullets were too small, and few of them took the rifling.

It fired with great rapidity, but soon got out of order, and would not be likely to remain long in proper trim . . . It might be useful in the defence of a narrow passage or bridge, but it is questionable whether it would be of any great practical utility in the open field of battle.[45]

Granted that the ammunition was defective one might have thought that the officers would have been able to extrapolate from what they had seen to imagine the effect such a weapon might have if it was using reliable cartridges. But some soldiers at least were prepared to persevere, and in 1867 the Gatling was officially tested by the Army. Two years later it was officially adopted for use. But because of the size of the gun, relative to the ordinary rifle, the military insisted upon regarding it as a piece of artillery, and thus mounted it upon large, fixed artillery carriages. This made it impossible to traverse the gun and thus considerably reduced its effectiveness. The same was true of the hand-cranked Gardner gun, adopted in 1881.

Two additional factors further complicated the problem. Firstly, as has already been noted, up until 1890 the artillery, at Woolwich, dominated British military thinking and they were most loath to concede that the machine gun was an infantry weapon and thus have it taken out of their own control. Both Chelmsford and Wolseley were totally opposed to this false classification. The former said in 1879: 'Machine guns should . . . in my opinion, not be attached to the artillery, but should be considered as essentially an infantry weapon.' In 1885 Wolseley was equally emphatic: 'There certainly is a very common impression in the minds of a great number that opposition from Woolwich has prevented our having . . . machine guns for many years past. But . . . this small arm is an infantry arm – not an artillery arm . . .'[46]

The second factor was the role of machine guns in the Franco-Prussian War. In the years prior to its outbreak in 1870 Napoleon III had given much encouragement to the development of a crank-operated machine gun known as the Montigny *mitrailleuse*. Work on it had been conducted in the utmost secrecy, at least as far as the French were concerned. In fact foreign military observers had been able to see the gun, and articles about it had appeared in several European military journals. But when the gun was triumphantly unveiled as the new secret weapon for the forthcoming war, hardly anybody in the French Army knew how to operate it. Worse still, the gun was cast in completely the wrong role. Like the British Gatlings and Gardners, it was mounted on an artillery carriage and assigned to conventional gunners. The name *'mitrailleuse'* itself tells the whole story for translated it means 'grape-shot firer'. And just before the beginning of the war the French artillery was reorganised into regiments of two six-gun batteries and a third battery of ten *mitrailleuses*. As Hutchison puts it: 'The organisation of the *mitrailleuses* was equivalent to a reduction of the French artillery by one third, and contributing the doubtful compensation of adding to each group of batteries a bundle of rifles, with the mechanical principles of which only a few were familiar.'[47] Naturally enough, the gunners, ignorant of the nature of the weapon that had been dropped into their laps, sited the machine guns with the ordinary field artillery. Equally naturally they were completely outranged by the Prussian Krupp pieces. At the crucial battles of Wissembourg and Spicheren they were blown to bits by the Prussians before they had a chance to fire.

From this experience it became clear to all observers that machine guns could be of little use if used as an adjunct to artillery. In fact, on some occasions they had been used to support the infantry and had been very effective in that role, as for example, at the Battle of Gravelotte in early 1871. But it suited no one to recognise such successes. Whether effective or not, the Woolwich group were determined that machine guns should remain with the artillery. Thus the successive failures in the Franco-Prussian War merely served for them to damn such weapons completely. As for the cavalry and the infantry, they were enmeshed in their obsolete conceptions of what war should be like, and this reactionary traditionalism had no place for the deployment of automatic fire that could wipe out whole companies at a time. The artillery was happy to hang on to the machine gun at any price, even if it meant making of them a tactical absurdity. The other arms were equally happy to surrender control as long as this meant that their effectiveness on the battlefield was almost nullified. The vested interests of both sides came together to ensure the continued neglect of the machine gun. Thus the artillery fixation lasted in Britain for a remarkably long time. Even after the introduction of the much lighter Maxim gun in 1887 few people saw that such a gun could become a vital mobile infantry weapon. Some soldiers actually wanted to retain the Gardner in preference to the Maxim, despite the fact that, as a military correspondent of the time wrote: 'The Maxim does not weigh more than forty pounds, while the Gardner weighs a hundred pounds; and as regards rapidity of fire and accuracy, the Maxim has the best of it, and is in all respects a far superior weapon.'[48] Even in the Boer War Maxims were mounted on a carriage that weighed four hundredweight with wheels that were a little under five foot in diameter. Usually they suffered the same fate as had befallen the *mitrailleuses* in the Franco-Prussian War. At the Battle of Paardeberg, for example, three guns and teams were destroyed in ten minutes.

From Maxim's point of view it must have been gratifying that the British Army had deigned to recognise the gun at all. His sales campaign in other countries had been going rather badly. Possibly the worst reaction of all had come from the Turks. The inventor had taken the gun to Constantinople in the hope of being able to give a demonstration. But on arriving at the War Ministry, so Maxim assures us, he was told by an English-speaking official: 'Hang your guns, we don't want your guns. Invent a new vice for us and we will receive you with open arms; that is what we want.' Even the inventive genius of Hiram Maxim was not up to that particular task. Nor, in these early days, was there much encouragement within his own firm. Sensibly enough he had taken on to his staff a military adviser, referred to throughout Maxim's autobiography as Captain X. It is probably for the best that his name was not passed down to posterity, for the advice he gave was worse than useless. Even in the context of the British Army at this time his views were somewhat extreme, particularly on the subject of armaments. When Maxim designed a non-recoiling field gun, Captain X was asked for his opinions. He declared himself opposed to the idea, and listed among his objections the fact that 'guns were not as a rule made for actual warfare, but for show', and that the gun I had designed was extremely ugly as compared with the graceful form of existing guns.'[49]

But the army as a whole was not much better. Though the first Maxims had been purchased in 1887, and by 1890 each battalion was supposed to have one gun for instructional purposes, in fact, right up until 1894, most of those soldiers who were taught about machine guns learned on outdated Nordenfelt and Gardner guns. Even in 1897 each battalion still did not have its own Maxim.

Then there came the event which of all others should have precipitated a drastic reappraisal of military thinking about machine guns. This was the Russo-Japanese War of 1904–5. Even though it took place on the other side of the world, this did not prevent most European countries from sending their own observers. Their eye-witness impressions and conclusions were widely disseminated in numerous newspapers and military journals, all of which were freely available to any British soldier who cared to learn something about the actual shape of modern warfare.

By far the most obvious factor was the devastating effect of firepower, and of all the weapons that contributed to it the machine gun was clearly seen to be the most deadly. Observer after observer noted this simple fact. The Russians, in fact, had been one of the first nations to realise something of the potential of machine guns. They had tested two Gatlings in 1865 and had promptly ordered twenty for the army. In 1868 they ordered a further seventy, and when they increased this order to one hundred guns the Russian Government was granted a license to manufacture Gatlings in Russia. These guns were known as Gorloffs. By 1876 the Russians had 400 Gatlings organised in eight-gun batteries. They were used quite extensively in the Russo-Turkish War of 1877–78. They were valuable additions to the defences at the sieges of Nicopolis and Plevna. The Russians at this time also regarded machine guns as an indispensable part of their cavalry formations. In the campaigns in Central Asia, particularly that in Khiva in 1870, such guns were used to deadly effect to break up the Turkish cavalry charges. In one offensive the whole Turkish offensive was broken up by the judicious use of two Gorloff guns.

With the introduction of fully automatic weapons the Russians at first proved a little obtuse. They simply could not conceive of a gun that was not crank-operated. When Maxim first visited Russia to give a demonstration he was asked what his gun's rate of fire was. His reply was greeted with derision on the gounds that it would be quite impossible for a man to turn the crank so rapidly. But unlike so many others the Russians were prepared to be convinced by the demonstration itself. By the time of the outbreak of the Russo-Japanese War their armies were well supplied with these guns. Each division had its own machine gun company, organised into batteries of eight Maxims. At the beginning of the war the Japanese were not nearly so well equipped, but they were not slow to learn the lesson. After a few months, all their infantry divisions, too, had a complement of 24 machine guns, though the Japanese favoured the Hotchkiss over the Maxim.

Both sides quickly found that machine guns could be of vital importance on the battlefield. A Russian divisional commander told the British military attaches that 'he was perfectly satisfied with his machine gun company, and that he thought machine guns absolutely necessary for infantry.' An attache with the Japanese wrote: 'On several occasions the Japanese left at the Battle of Shou-shanpou were checked by machine gun and rifle-fire, and there is no doubt that a strong feeling exists in the infantry that the presence of machine guns with the Russian Army confers upon it a distinct advantage.' Of the Russian machine guns used in the defence of Port Arthur, a British war correspondent wrote: 'The death-dealing machine guns of the Russians in the casements of the fort are playing ghastly havoc – such havoc that only a score or more of Ouichi's battalions reached the first ditch of the defence, where they threw themselves panting into the pits their own artillery had torn.' Another correspondent said of these same guns: 'Nothing can stand against them, and it is no wonder that the Japanese fear them, and even the bravest have a chilly feeling creeping down their backs when the enemy's machine guns beat their devil's tattoo.'[50] A German correspondent with the Russians saw an example of just why this should be so:

On January 8, 1905, near Lin-chin-pu, the Japanese attacked a Russian redoubt armed with two Maxim guns. A Japanese company, about two hundred strong, was thrown forward in skirmishing order. The Russians held their fire until the range was only three hundred yards; the two machine guns were then brought into action. In less than two minutes they fired about a thousand rounds, and the Japanese firing-line was literally swept away.[51]

Captured Russian Maxims after the Battle of the Yalu 1904

But once the Japanese got their own machine guns they were able to use them equally effectively. A correspondent from the *Kolnische Zeitung* wrote: 'In the offensive the Japanese frequently made effective use of machine guns. When the infantry were carrying out a decisive attack, they were supported by their machine guns, which concentrated their fire on points arranged beforehand . . . When machine guns have been skilfully employed their action has been infinitely more effective than that of the field artillery . . .'[52] A German military attache wrote: 'At Mukden . . . all the machine guns of a whole Japanese division were brought into action upon a Russian *point d'appui*. The Russian fire was silenced, but burst out again whenever the machine gun fire slackened. The Japanese infantry used these pauses in the enemy's fire to press forward to close range under cover of their machine gun fire.'[53] Summing up the whole experience of the war, the German Official History noted that 'machine guns were extraordinarily successful. In defence of entrenchments especially they had the most telling effect on the assailants at the moment of assault.'[54]

The lessons to be drawn were obvious. As has been seen already, the Germans set to work immediately and began to develop a much larger machine gun capability. The French followed suit. But British would-be reformers were still hamstrung by the traditionalism and myopia of the incumbent High Command. Perhaps there is some slight excuse for not being responsive to possible inferences from a war waged on the other side of the world. But one might have thought that, in the context of the great European arms race of the early twentieth century, those in power would have looked upon German rearmament in this sphere with some apprehension. As early as 1887, Lord Beresford had noted that 'it would be a very serious thing if the German or any other army were to take up the machine gun question. Whilst we, with all our practical experiences having found it so useful on so many occasions, were not to take it up and thrash it out . . .'[55] In 1911 Repington also tried to draw attention to the emphasis which the German Army was now placing on the use of machine guns. As he said: 'Nations which may have to encounter the German Army must strive to excel in the use of this weapon, which proved its worth in Manchuria.'[56]

But the British military paid no heed to these warnings. In 1914 each infantry battalion still only had an allotment of two Vickers guns (almost identical to the Maxim) each. On top of this, one of the guns 'was tucked away for special purposes and the other was used for training and exercises. The result was that stress was inevitably laid in the role of a single gun generally in a colonial and minor context.'[57] With only the one gun per battalion, and the paucity of machine gun courses at Hythe, the inevitable result was that hardly anyone in the battalion actually knew how to operate a machine gun. When the British Expeditionary Force arrived in France in 1914 it was found that in many units only one or two subalterns had any knowledge of the automatic mechanisms. One authority has admirably summed up the pre-war attitude. When a battalion or a division was on manoeuvres a junior officer would inevitably ask, 'What shall I do with the machine guns today, sir?' The equally inevitable reply would be, 'Take the damned things to a flank and hide them.'[58]

More recently, another writer has tried to excuse this continued neglect of the machine gun:

> The machine guns in use were clumsy, too large to be manhandled for more than short distances, and were water-cooled, with delicate mechanisms that often jammed under the rough conditions of the battle-field . . . Rapid fire by trained riflemen was more reliable, and remained so until the unexpected trench warfare of 1915 gave the machine gun its opportunity to dominate the battlefield. Nevertheless, the army had recognised the machine gun as the weapon of the future and had approached the Treasury for an increased allotment of money to provide machine guns. It was when their requirement was flatly rejected in 1907 that the generals decided upon the alternative of making British soldiers the best riflemen in the world.[59]

But such an explanation is wrong on almost every count. There was little need to manhandle the machine guns at all because their correct role was one of giving fixed fire from behind the lines. The mobile light machine guns did not appear until later in the war: It is also a little unfair to make such a point of the delicacy of the mechanism itself. Such a remark is more appropriate to the days of the crank-operated Gatlings and Gardners which were certainly prone to jam almost every time they fired. But the machine guns of 1914 were much more reliable affairs. One might recall, for example, that at trials of an early Maxim model, at Spezzia, in Italy, the gun had been submerged in the sea for three days before, without being cleaned – it gave a faultless performance. One might also remember that 'before the war, a machine gunner was trained to tap the butt of his gun with sufficient strength to move it two inches, which deflected the muzzle so that at . . . two hundred yards there was an unbroken arc of fire.'[60] This 'tap' does not seem to have had any effect on the gun's ability to fire. And even granted that rapid fire by trained riflemen was more reliable, the essential point is that it took time to train these riflemen. If ever conscription were necessary the average recruit would be a poor second best to any machine gun, however unreliable. And why was trench warfare so unexpected in 1915? This whole chapter has been devoted to showing why such a form of warfare was almost inevitable. Nor can one claim that the far-sighted generals were baulked by parsimonious bureaucrats. In 1908 the Financial Secretary at the War Office, impressed by official observers' report about the German Army, had written to the Master-General of Ordnance to say: 'If the military members of the Council would like to have more machine guns for the Army . . . the Finance Department of the War Office would have no objection.'[61] He received the reply that two per battalion were considered quite sufficient. Finally, it was not the generals who decided to fall back on the highly trained rifleman, it was those in charge of musketry training, who were forced into this expedient by the short-sightedness of the generals themselves.

For military reactions to the machine gun were not a rational response to either technical or financial considerations. They were rooted in the traditions of an anachronistic officer corps whose conception of warfare still centred around the notions of hand-to-hand combat and individual heroism. They still thought that they would fight on the battlefields on which man was the dominant factor and only needed courage and resolution to be able to carry the day. To have the good fortune to be British, or French or German, meant that by definition one also had the good fortune to possess such characteristics. Such men could not admit that a mere machine had robbed them of their old primacy. Thus they ignored it. For the British commanders on the eve of the First World War the machine gun simply did not exist.

This obdurate attitude was also noticeable outside the army. In a series of articles written in the *Manchester Guardian* just after the outbreak of war that paper completely underestimated the significance of the machine gun, obviously basing their assessment on Britain's colonial experience: 'The machine of today, which may be described as an automatic rifle on wheels, fires some hundred shots a minute, but it is not useful at long ranges or for what is called a "shallow target", that is, for example, a single line of men in a trench. It is deadly against a "deep target", such as a facing column of men at fairly close range.'[62] In *Nash's War Manual*, written to explain to civilians the balance of forces at the beginning of the war, machine guns are not even mentioned. This is hardly surprising given the author's basic premise that: 'The armies in the field today are mobile beyond the wildest dreams of strategists of a former generation.' Yet the author does find time to go into raptures about a new eight-pound mine grenade 'containing four hundred large bullets . . . When the enemy is over the mine the touch of an electric button causes it to spring out of the ground until it is checked by a chain at the height of a yard above the surface, when it explodes, mowing down every man in the vicinity.'[63] Only a complete misunderstanding of the nature of modern warfare could prevent someone from seeing that in comparison to the machine gun such a device was mere child's play.

Military resistance to the machine gun was not only confined to Europe. Surprisingly, perhaps, the American Army too completely missed the potential importance of such weapons. Civilian interest in automatic fire first emerged in America because it was there that people had first discerned the shape of industrial mass society, and had first had a glimpse of the nature of mass warfare. But such prescience was confined to the inventors. The soldiers themselves consistently failed to realise that automatic weapons were now a fundamental component of a modern armoury. The basic reason for this apparent contradiction was that Americans, with traditional mistrust of military establishments, have always, or at least until 1945, regarded war as an unfortunate aberration. As soon as any conflict in which the Americans were engaged was over they would begin to demobilise as swiftly and as completely as possible. In 1783 and 1865, not to mention 1918 and 1945, the American Army was almost immediately reduced to a fraction of its wartime strength. After each war all that was left was a rump of professional, long-serving regulars. Thus, no matter what the scale and character of the actual wars in which America participated, for the bulk of the time her army was very small and the men in it, though they did not have the same aristocratic traditions as in Europe, were similarly cut off from the march of events in the world outside.

Machine guns did have some supporters but they were few and far between. In 1862 the *Indianapolis Gazette* came out in emphatic support of the Gatling gun:

> They . . . can be made so important at the critical moment – the turning point in a battle – that it does seem strange to us that the Government has not long since ordered a large number of them to be made for use of the army. Every regiment ought to have at least one of them, and it would be well in some cases if every

company had one. In cases of battle they should probably only be used as a reserve, to be brought up and
turned on the enemy at the critical moment, or in case
the enemy is making a charge. Three or four of them in
such a case would be equal to as many fresh regiments,
with not one-tenth the danger of loss of life on our side.

A few military men were also impressed by the Gatling's
potential. In 1866 a captain of Ordnance wrote to say that he
had found the gun very satisfactory: 'The moral effect of the
Gatling gun would be very great in repelling an assault, as
there is not a second of time for the assailants to advance
between the discharges . . . In my opinion, this gun could be
used to advantage in the military service.'

But after the cessation of hostilities in 1865 few military
men bothered to even acknowledge the existence of the
machine gun. One of these was General Norton who in 1882
made a very far-sighted proposal:

> The startling proposition . . . suggests itself to do
> most if not all our long range firing with Gatling bat-
> teries. In future wars this fire must systematically be
> resorted to . . . It means of course to increase our
> number of Gatling guns until we have a Gatling corps
> as numerous as artillery itself. The first expense is of
> small moment comparable with promises and results of
> warlike preparations; and this method promises not
> only double the effect at half the transportation and
> expenditure of ammunition, but will enable us to keep
> our infantry well in hand, and more or less out of the
> battle until well within short range and decisive dis-
> tances.

No effort at all was made to act upon these proposals.

The Americans were actually given one more chance to
see the effectiveness of machine guns in action. During the
Spanish-American War one of their greatest advocates,
Lieutenant J.H.Parker, managed to snatch the opportunity
to take a small number of Gatlings to Cuba to support the
American troops there. At the Battle of Santiago, during the
assault on San Juan Hill, they did sterling service in supporting the American attacks. For Parker this was final vindica-
tion, and in 1899 he wrote three books singing the praises of
such weapons. For him: 'All that had ever been claimed by
the most enthusiastic weapons was here exemplified, and the
result of the campaign was to place the machine gun beyond
dispute as a weapon to be reckoned with in some form in all
future wars.'[64] Elsewhere Parker's attitude is very reminis-

cent of the elitism that is to be found in the German Field Regulations or the writings of Guerin and Lavau. Thus:

> The machine gun man must be hot-blooded and dashing. He must have all the nerve and *élan* of the best light cavalry, all the resisting power of stolid and immovable infantry. He is not to reason abstruse theorems, nor approximate different ranges; his part is to dash into the hell of musketry, the storm of battle, and to rule that storm by the superior rapidity and accuracy of his fire.[65]

Parker was not the only person to take note of the important role of the machine guns in Cuba. Theodore Roosevelt, in his account of his days with the Rough Riders, noted that at one stage in the battle of Santiago the cry went up, ' "Its the Gatlings, men, our Gatlings!" It was the only sound which I ever heard my men cheer in battle.'[65]

But neither humble lieutenants nor eminent public figures could make any impact on the conservative military establishment. From the very beginning the machine gun was almost completely absent from its calculations. Machine guns, for example were hardly ever used in the campaigns

American Gatlings at San Juan Hill 1898

against the Indians. Even when they were available commanders seemed loath to actually take them on active service. General George Armstrong Custer was guilty of monumental folly in this respect. In 1876, when he led his entire troop to be massacred at the Battle of the Little Big Horn, there were four Gatlings available with his headquarters. But Custer declined to take any of them with him. One can only explain this oversight as a typical piece of arrogant bravado. He said that it would be too difficult to haul them over the terrain, but in fact they were of a model specially designed to be dismantled and carried on pack mules.

Much of the blame for the subsequent neglect of the machine gun must be laid at the door of the Ordnance Department. The reactionary attitude of Colonel Ripley during the Civil War has already been mentioned. His policy was consistently followed right up until 1917. Over the years the Department built up a proud tradition of totally ignoring the most significant developments in machine gun technology. The Maxim gun, the Browning medium machine gun, the Lewis gun and the Browning automatic rifle were all tested and rejected by the American Army's firearms experts. Without the intervention of the First World War it is extremely doubtful whether the United States would have had a machine gun capacity that was more than a fraction of that of the European powers.

Possibly the most inexplicable rejection was that of the Lewis gun in 1913. There has been much controversy surrounding this decision. The Ordnance Department alleged that the gun was subject to frequent breakdowns. More recently it has been claimed that the recoil mechanism in the early prototypes was not strong enough to withstand the constant pressure put upon it. What does seem true is that there was a serious clash of personality between Colonel Lewis, the inventor, and General Crozier, the head of the Ordnance Department. In many ways this is reminiscent of Ripley's mistrust of Gatling. As one writer has noted:

> Defence of the Ordnance Department's policy at times has taken the form of criticism of the attitude or activities of Colonel Lewis, the implication being that he deserved no further consideration. This suggests something of the same attitude that prevailed at times during the Civil War – that the treatment accorded new weapons was related to the personality or the deserving character of the inventor.[67]

It is certainly true that, even if the gun did not function well at its first trial, the subsequent whole-hearted espousal of this weapon by most of the major Allied powers might have persuaded Crozier and his subordinates to try again. As a British correspondent wrote in the *Philadelphia Public Ledger* in February 1917:

> I do know that . . . some 4,000 officers and about 40,000 men use the Lewis gun . . .exclusively, and that in the British, French, Italian and Russian armies there are at this moment nearly 40,000 in actual and daily operation : . . That after fully two years of daily experience in the battlefields it stands higher than ever in the judgement of the British armies. Yet this is the gun the American Government . . . turned down. I have heard all sorts of explanation of its action, mainly of a personal or political character. But I have never yet heard it asserted that the Lewis machine gun was rejected by the authorities at Washington on its merits or that they have any better gun or any that is as good up their sleeve. Incidents such as these have a somewhat more than depressing effect on an Englishman who has seen at first hand the terrible effects of a state of unpreparedness and who has no dearer wish than that the United States may be wise in time.

Even the British Army itself, after two years exposure to the absolute necessity of maintaining an adequate force of machine guns, were a little bewildered by the American refusal to give the Lewis gun another chance. In October 1916 a writer in the Cleveland Press noted that:

> Nothing the whole war has brought out has been of so much real use to the Brtitish Army as the Lewis machine gun. It has done wonders. It has almost counteracted the British aversion to tactics . . . One odd little fact is that they nearly all believe the American army to be equipped with the Lewis gun. 'But then you have your Lewises', British Army officers would say to me when the United States was trying to get an army to the Mexican border. 'Rum country, rum country,' they would say to me when I told them we hadn't.

There were even some members of the American military who thought that their country might do well to adopt the Lewis gun. The General in command of the Department of the East was quite definite that, 'in my private opinion, the Lewis machine gun is the best light-type gun yet developed for troops in the field . . . We need a reserve supply of 25,000 machine guns as in the end one in ten men will carry a 26-lb machine gun as he now carries a rifle.'[68]

But none of this had any impact on the higher authorities. The Lewis gun continued to be ignored. Even when the American Navy decided that the gun was quite adequate and equipped the Marine Divisions with it, the army remained adamant. When these divisions arrived in France they were attached to army units and put under the command of the latter. As soon as this happened they were ordered to turn in their Lewis guns and be re-equipped with the notorious Chauchat.

So on American entry into the First World War they found that they had at their disposal the grand total of 1,453 assorted machine guns. By far the biggest percentage of these were the Benet-Mercié machine gun of 1909. Though reasonably reliable the gun did have some unfortunate drawbacks. It was used in 1916 during the border trouble with Pancho Villa. Yet during a Mexican night raid on Columbus, New Mexico, the American troops were unable to bring their machine guns into action because the gun's system of loading was so intricate that they could not be used after dark. Why the Americans should have chosen to adopt such a gun, in preference to far superior models, must remain a mystery. However, one might note in passing that Laurence V. Benet, its co-developer, was the son of General S.V. Benet, a past Chief of Ordnance. Clearly the gun itself, and the numbers in which it was available, were totally inadequate to American needs. A desperate last-ditch effort was made to produce reliable machine guns in sufficient quantities. American factories were retooled and vast numbers of Vickers and Browning guns were produced. Between July and September 1918 the average monthly output of machine guns of all types was 27,270 as against 12,126 in France and 10,947 in Britain.

Unfortunately these weapons did not arrive in France in any significant quantity until the war was over. In the meantime the Americans were forced to rely upon the Chauchat, which, as has been seen, the French chose to regard as a machine gun and the Americans threw away as so much scrap metal. Nor did the American High Command help. One has already seen how the Marines were deprived of their Lewises. When the first Browning automatic rifles arrived these too were withdrawn from service. Because they were so efficient, General Pershing feared that the Germans might capture one and copy the design.

Notes

1. P.G.Razzell, 'Social Origins of Army Officers', *British Journal of Sociology*, vol.14, no.3, 1963, p.253.
2. P.M.de la Gorce (trans. K.Douglas), *The French Army*, Weidenfeld and Nicolson, London, 1963, p.35.
3. A.Vagts, *A History of Militarism*, The Free Press, New York, 1967, p.224.
4. Ibid., p.156.
5. Ibid., p.195.
6. Ibid., p.213.
7. I.F.Clarke, *Voices Prophesying War, 1763–1984*, Panther Books, London, pp.89–90.
8. Ibid., p.127.
9. For a good brief summary of these developments see C.Falls, *A Hundred Years of War*, Duckworth and Co., London, 1953, Chapter 4.
10. J.Luvaas, *The Military Legacy of the Civil War*, University of Chicago Press, Chicago, 1959, p.107.
11. Luvaas, *The Education of an Army*, Cassell, London, 1965, p.114.
12. Luvaas, *Military Legacy*, op.cit., p.56.
13. Clarke, op.cit., p.134.
14. Luvaas, *Education*, op.cit., pp.181 and 212.
15. Ibid., p.243.
16. Luvaas, *Military Legacy*, op.cit., p.17.
17. S.Wilkinson, *War and Policy*, Dodd, Mead and Co., New York, 1900, p.159.
18. Col.R.Meinertzhagen, *Army Diary, 1899–1920*, Oliver and Boyd, Edinburgh, 1960, p.15.
19. J.F.C.Fuller, *Armaments and History*, Eyre and Spottiswoode, London, 1946. p.135.
20. Luvaas, *Military Legacy*, op.cit., pp.5. and 28.
21. B.H.Liddell-Hart, *Foch: Man of Orleans*, Penguin Books, Harmondsworth, 1937, vol.1, p.76.
22. Kranzberger and Pursell, op.cit., p.499.
23. Luvaas, *Military Legacy*, op.cit., pp.28 and 5.
24. Ibid., p.111.
25. Meinertzhagen, op.cit., p.8.
26. B.Bond, *The Victorian Army and the Staff College 1854–1914*, Eyre Methuen, London, 1972, p.186.
27. Bond, 'Doctrine and Training in the British Cavalry, in M.Howard (ed.), *The Theory and Practice of War*, Cassell, London, 1965, p.117.
28. Luvaas, *Military Legacy*, op.cit., p.198.
29. B.H.Liddell-Hart, *The Tanks*, Cassell, London, 1959, vol.1, p.234.
30. Luvaas, *Education*, op.cit., pp.300 and 316.
31. Bond, 'Doctrine and Training', op.cit., p.120.
32. Hutchison, op.cit., p.39.
33. Wahl and Toppel, op.cit., p.105.
34. Hutchison, op.cit., p.47 and J.Symons, *England's Pride, The Story of the Gordon Relief Expedition*, Hamish Hamilton, London, 1965, p.275.

35. J.Lehmann, *All Sir Garnet*, Jonathan Cape, London, 1964, p.283.
36. Hutchison, op. cit., p.72.
37. C.H.B.Pridham, *Superiority of Fire*, Hutchinson, London, 1945, pp.54 and 59.
38. Hutchison, op.cit., p.36.
39. B.Gardner, *Allenby*, Cassell, London, 1965, p.63.
40. Meinertzhagen, op.cit., p.55.
41. Bond, *Staff College*, op.cit., p.291.
42. Maxim, op.cit., p.210.
43. F.V.Longstaff and A.H.Atteridge, *The Book of the Machine Gun*, Hugh Rees, London, 1917, p.156.
44. Ibid., pp.74 and 84.
45. Luvaas, *Military Legacy*, op.cit., p.25.
46. Hutchison, op.cit., pp.39 and 47.
47. Ibid. p.26.
48. Pridham, op.cit., p.41.
49. Maxim, op.cit., pp.230 and 205.
50. Hutchison, op.cit., pp.84, 89 and 90.
51. Pridham, op.cit., p.47.
52. Hutchison, op.cit., pp.92–3.
53. Hobart, op.cit., p.109.
54. Pridham, op.cit., p.49.
55. Ibid., p.31.
56. Luvaas, *Education*, op.cit., p.315.
57. W.Moore, *See How They Ran*, Leo Cooper, London, 1970, p.36.
58. C.D.Baker-Carr, *From Chauffeur to Brigadier*, Ernest Benn, London, 1930, p.80.
59. C.Carrington, *Soldier From the Wars Returning*, Hutchinson, London, 1965, p.25.
60. G.Chapman, *A Passionate Prodigality*, Macgibbon and Kee, London, 1965, p.42.
61. B.H.Liddell-Hart, *History of the First World War*, Pan Books, London, 1973, pp.142–3.
62. Manchester Guardian, *Weapons of War*, John Heywood, Manchester, 1914, p.30.
63. *Nash's War Manual*, Eveleigh Nash, London, 1914, p.275.
64. Wahl and Toppel, op.cit., pp.19, 101 and 143.
65. Hutchison, op.cit., p.73.
66. Fuller, op. cit., p.131.
67. J.A.Huston, *The Sinews of War: Army Logistics, 1775–1953*, Office of the Chief of Military History, Washington, 1966, p.322.
68. Chinn, op.cit., vol.1, pp.288–9.

IV *Making the Map Red*

'And the white man had come again with his guns that spat bullets as the heavens sometimes spit hail, and who were the naked Matabele to stand up against these guns?'

A great gulf existed between the effectiveness of military firepower and the soldiers' total lack of respect for its potential. Their contemptuous attitude persisted, even in the teeth of mounting evidence of the unparalleled efficacy of modern firearms. Nor had that evidence been manifested only in parts of the world not very familiar to European military establishments. The machine gun had also been put to use in an area of the world with which most European armies had close connections, and there its effectiveness had become shatteringly obvious. In Africa automatic weapons were used to support the seizure of millions of square miles of land and to discipline those unfortunates who wished to eschew the benefits of European civilisation. With machine guns in their armoury, mere handfuls of white men, plunderers and visionaries, civilians and soldiers, were able to scoff at the objections of the Africans themselves and impose their rule upon a whole continent.

Without examining all the reasons for imperialist expansion it is certain that the search for markets, strategical considerations and the question of national prestige were all contributory factors, though historians have argued about the exact importance of each. But of one thing there is no doubt. Whatever the general causes, or the personal motives of the individual colonisers, the whole ethos of the imperialist drive was predicated upon racialism. Attitudes to the Africans varied from patronising paternalism to contempt and outright hatred, but all assumed that the white man was inherently superior to the black.

A central strand of this racialism was the crude interpretation of Darwinian theories about 'the survival of the fittest'. Projected back into the past, such theories enabled people to put forward the relative superiority of Western civilisation as a reason for arguing that the white man was the dominant race. One could also extrapolate from them to predict that eventually this race would physically dominate the whole world. An extreme version of this prediction was given in 1881, by W.D.Hay, in a book called *Three Hundred Years Hence*. He described the future paradise:

> The old idea of universal fraternity had worn itself out; or rather it had become modified when elevated into the practical law of life. Throughout the Century of Peace . . . men's minds had become opened to the truth, had become sensible of the diversity of species, had become conscious of Nature's law of development . . . The stern logic of facts proclaimed the Negro and Chinaman below the level of the Caucasian, and incapacitated from advance towards his intellectual standard. To the development of the White Man, the Black Man and the Yellow must ever remain inferior, and as the former raised itself higher and yet higher, so did these latter seem to sink out of humanity and appear nearer and nearer to the brutes . . . It was now incontrovertible that the faculty of Reason was not possessed by them in the same degree as the White Man, nor could it be developed by them beyond a very low point. This was the essential difference that proved the worthlessness of the Inferior Races as contrasted with ourselves, and that therefore placed them outside the pale of Humanity and its brotherhood.

Clearly, working from such a theory of human development, it was easy, even natural, to go on to regard superior military technology as a God-given gift for the suppression of these inferior races. A popular history of science of 1876 offers a perfect example of such an attitude. In the chapter on firearms the author tells us:

> We often hear people regretting that so much atten-tion and ingenuity as are shown by the weapons of the present day should have been expended upon instru-ments of destruction . . . The wise and the good have in all ages looked forward to a time when sword and spear shall be everywhere finally superseded by the ploughshare and the reaping-hook . . . Until that happy time arrives . . . we may consider that the more costly and ingenious and complicated the instruments of war become, the more certain will be the extension and the permanence of civilisation. The great cost of such appliances as those we are about to describe, the ingenuity needed for their contrivance, the elaborate machinery required for their construction, and the skill implied in their use, are such that these weapons can never be the arms of other than wealthy and intelligent nations. We know that in ancient times opulent and civilised communities could hardly defend themselves against poor and barbarous races . . . In our day it is the poor and barbarous tribes who are everywhere at the mercy of the wealthy and cultivated nations.[1]

The Europeans had superior weapons because they were the superior race. With regard to the machine gun, for example, one writer assured his readers that 'the tide of invention

Gatling practice during the Zulu War 1897

which has . . . developed the "infernal machine" of Fieschi into the mitrailleur (sic) and Gatling Battery of our own day – this stream took its rise in the God-like quality of reason.'[2] Thus when the Europeans opened their bloody dialogue with the tribes of Africa it was only natural that they should make them see reason through the ineluctable logic of automatic fire.

The British Army had decided to purchase twelve Gatlings in October 1869 but they were not sent on active service until 1874, on the occasion of the first campaign against the Ashantis. It was decided that Wolseley's small expeditionary force should take along some Gatlings to even up the odds. *The Times* heard of this decision in late 1873 and it prompted them to express some rather bloodthirsty hopes:

> The Gatling guns . . . we presume are mainly intended for the defence of stockaded positions. For fighting in the bush a Gatling would be as much use as a fire engine, but if by any lucky chance Sir Garnet Wolseley manages to catch a good mob of savages in the open, and at a moderate distance, he cannot do any better than treat them to a little Gatling music . . . Altogether we cannot wish the Ashantees worse luck than to get in the way of a Gatling well served . . .[3]

Wolseley himself had prophesied to his soldiers that the Gatlings would provoke a feeling of 'superstitious dread' in the natives. Eager to take early advantage of such moral factors, he staged a demonstration of one of the Gatlings before a group of tribesmen. Unfortunately the gun immediately jammed, leaving the natives with a rather over-optimistic assessment of British firepower. When the time came to press on into the interior, Wolseley, probably aggrieved at this embarassing failure, decided not to take the Gatlings along, claiming that the terrain was quite unsuitable.

This was a rather inauspicious start to the history of machine guns in Africa, but it was one of very few times that they were to prove such a disappointment. Their next trial was much more promising. In 1879 two successive expeditions, under the command of Lord Chelmsford, were sent into Zululand against the *impis* of the great Cetshwayo. The first met a most humiliating reverse at the Battle of Isandhlwana, and was also the occasion of the unnecessary heroics of Rorke's Drift. This latter engagement had convincingly displayed the power of modern musketry when pitted against mere hide shields and *assegais*, but on the second expedition Chelmsford was determined to put the issue quite beyond doubt, and four Gatlings accompanied the troops. The Navy had actually supplied one Gatling and crew for the first campaign, but this had been placed with Colonel Pearson's flank column on the right and it had been unable to make any contribution to the main engagements. Nevertheless, it had been well served. It was in the charge of Midshipman Lewis Cadwallader Coker whose dedication would have delighted those, like Parker or Guerin, who looked on machine gunners as an *élite* corps. Coker died of disease whilst in the field. His chances of recovery had been minimised because 'to the last he had insisted on sleeping in the open beside his beloved Gatling gun.' Similar dedication was shown by Midshipman Morehead during the Battle of Ulundi, the climax of the second campaign. At one stage in the fighting he 'was hit in the thigh, pouring blood over the frame that held the chattering barrels. He waved the litter-bearers aside, and sank down beside the gun to help load the drums of ammunition.'[4]

Artist's impression of a naval Gatling

This time such singlemindedness was not in vain. As Chelmsford mildly put it: 'At Ulundi we also had two Gatlings in the centre of the front face of our square. They jammed several times when in action but proved a very valuable addition to the strength of our defence.'[5] In choosing to fight in square formation Chelmsford had shown a typically aristocratic concern with outmoded notions of honour and fair play. Explaining his tactics to his officers, he insisted that 'we must show them that we can beat them in a fair fight.' Given the weapons at his disposal such chivalric niceties were a little irrelevant. By no stretch of the imagination could the Battle of Ulundi be deemed a fair fight. British casualties were slight, whilst, as Cetshwayo is reputed to have said after the battle, 'There are not tears enough to grieve for all our dead.' Not a little of the credit for this slaughter went to the Gatlings. As a war correspondent for the *London Standard* said: 'When all was over and we counted the dead, there lay, within a radius of five hundred yards, 473 Zulus. They lay in groups, in some places, of fourteen to thirty dead, mowed down by the fire of the Gatlings, which tells upon them more than the fire of the rifles.'[6]

The last important engagement in which Gatlings took part, again with the Naval Brigade, was the assault on Tel-el-Kebir in 1882. The *Army and Navy Gazette*'s account of this action sums up both the efficacy of the guns and the contempt for the enemy which made it so easy to use them:

> The naval machine gun battery, consisting of six Gatlings . . . reached the position assigned to it . . . Having received orders to advance they came within easy reach of the Tel-el-Kebir earthworks . . . The order 'action-front' was given and taken up joyously by every gun's crew. Round whisked the Gatlings, r-r-r-r-r-rum, r-r-r-r-r-rum, r-r-r-r-r-rum! that hellish note the soldier so much detests in action, not for what it has done, so much, as for what it could do, rattled out. The report of the machine guns as they rattle away rings out clearly on the morning air. The parapets are swept. The embrasures are literally plugged with bullets. The flashes cease to come from them. With a cheer the blue-jackets double over the dam, and dash over the parapet, only just in time to find their enemy in full retreat. That machine gun was too much for them. Skulking under the parapet they found a few poor devils, too frightened to retire, yet willing enough to stab a Christian, if helpless and wounded.[7]

The Naval Brigade also took part in the next expedition to

Egypt, the attempts to relieve Gordon in Khartoum in 1884 and 1885. There were several Gatlings in the gunboats that accompanied the British troops, but the only one to see service on land was a single Gardner gun with a small party of sailors under the command of Charles Beresford. They were accompanying a small column of soldiers that was surprised by the Arabs at Abu Klea. The British immediately formed a square with the Gardner in the middle and managed to fight off the Dervish assaults. In fact the Gardner only managed to fire seventy rounds before jamming, but even so its effect was most heartening. As Beresford himself said: 'As I fired I saw the enemy mowed down in rows, dropping like nine-pins.' And once again the men of the Naval Brigade showed themselves capable of great heroism in defence of their machine guns. At one stage in the battle Gardner actually moved the Gardner outside the square, thus helping the Dervishes to break in temporarily. In the hand-to-hand fighting that followed eight of his subordinates, the entire complement of the naval party, were killed as they successively tried to get the gun working again. Indeed Beresford's own attachment to the gun seems to have been a little excessive. In the later attempt to relieve Metemneh the British column was accompanied by a battery of seven-pounders 'but not the Gardner, possibly because Beresford was incapacitated by a painful boil on his bottom. One has at times the impression that Beresford regarded the Gardner as his personal property.'[8]

Gatlings at El Teb, in the Sudan, 1884

The climax of the wars in Egypt and the Sudan came in 1898 when the British Government decided to finish the

matter once and for all. But by this time their forces had the Maxim gun whose reliability was beyond question, at their disposal. The final showdown of the campaign came at the Battle of Omdurman where the bulk of the Dervish forces repeatedly hurled themselves against the British lines, and were repeatedly beaten back by the deadly small-arms fire. The Maxims were the most deadly component of this massed firepower. A German war correspondent with the British wrote: 'The gunners did not get the range at once, but as soon as they found it, the enemy went down in heaps, and it was evident that the six Maxim guns were doing a large share of the work in repelling the Dervish rush.'[9] Another eye-witness wrote of the effects of these weapons when he described the battlefield at the end of the day:

> It was the last day of Mahdism and the greatest. They could never get near and they refused to hold back . . . It was not a battle but an execution . . . The bodies were not in heaps – bodies hardly ever are; but they spread evenly over acres and acres. Some lay very composedly with their slippers placed under their heads for a last pillow; some knelt, cut short in the middle of a last prayer. Others were torn to pieces . . .[10]

Because Winston Churchill was a participant in this battle, contemporary mythology has retained nothing of it except the futile charge of the 21st Lancers, in which Churchill took part. But at this time a much more accurate assessment of the significance of Omdurman was made by Sir Edward Arnold. Maxim proudly quotes the following remark in his autobiography: 'In most of our wars it has been the dash, the skill, and the bravery of our officers and men that have won the day, but in this case the battle was won by a quiet scientific gentleman living in Kent.'[11] When one looks at the

Maxims at the Battle of Omdurman 1898

casualty figures for Omdurman, 28 British and 20 others killed against 11,000 Dervish *dead*, one can hardly ascribe to Sir Edward any particularly outstanding powers of analysis. Yet he was almost the only public figure who was prepared to admit that automatic firepower had won the day. For others it became another example of the triumph of the British spirit, and the general superiority of the white man.

In fact Maxims had already proved their worth in Africa before the Omdurman campaign. Further into the hinterland they had been used with unflagging zeal by various

Cleaning the Streets of Alexandria 1882

pioneers. For men like Cecil Rhodes and Frederick Lugard, and organisations like the British South Africa Company and the Imperial East African Trading Company, the Maxim gun was an indispensable tool for the imposition of European control. Maxim himself was only too pleased that his invention should be used in such a role. In 1887 an expedition led by Stanley set off for Wadelai, near Lake Albert, to rescue Emin Pasha (Eduard Schnitzer), who had established a bizarre dominion among the natives of that region. This expedition of mercy attracted the imagination of Europe and Maxim donated one of his guns to help them on their way.

In 1890 the somewhat battered gun was taken up by Lugard when he left Mombasa to travel to Uganda. By the Anglo-German Treaty of that year Uganda had been recognised as falling within the British sphere of influence, and Lugard lost no time in revealing the reality of that influence. Missionaries had already exacerbated tribal tensions there and relations between Protestant and Catholic Africans became increasingly bitter. In 1892 open warfare broke out, with the Catholics demanding the expulsion of the British. Lugard immediately threw his support, which included his Maxims, behind the Protestant Ingleza tribe. This support was decisive, despite Catholic over-optimism. During the preparation for the uprising, Mwanga, its leader, had drawn some faulty conclusions from the tatty appearance of Lugard's much-travelled Maxim. One of his envoys 'had circulated the most extraordinary reports, saying that we were cowards who dare not fight . . . that our Maxim was merely for show, and fired single bullets like a gun.' Clearly the usual efficacy of such weapons had already been noised abroad in Africa. But even a battered Maxim was better than no Maxim at all. At the Battle of Mango Hill Lugard and his Sudanese mercenaries threw their weight behind the Protestants, the former taking charge of the machine gun: 'Firing the Maxim hastily, Lugard scored a pair of freak hits on the legs of two Fransa chiefs . . . He then traversed to cover an open potato patch that the attackers would have to cross. The Maxim was now jamming at almost every other shot . . . but the few rounds Lugard got off sufficed to hold the Fransa back.'[12]

Perhaps Mwanga's original concern with the potential of the Maxim gun had been aroused by events in Tanganyika. In 1890 certain German opportunists had established the German East African Company. They almost immediately encountered African opposition and in 1891 the Company was involved in a savage war with the Hehe tribe under their chief Mkwawa. At one stage Hehe warriors had ambushed a German column and massacred almost everyone. But it turned out to be a Pyrrhic victory at best. Towards the end of the battle, 'the German officer-surgeon, helped by *askaris*, dragged two machine guns, with plenty of ammunition into a mud hut, and from there turned the tables on the Hehe. He is said to have killed about one thousand of them.'[13]

Probably the leading practical proponent of British imperialist expansion was Cecil Rhodes. By the same token he was keenly aware of the indispensable role of the machine gun in translating his visions into reality. In 1893 he went in person to Pondoland to meet Sicgau, a dissident chief. The bulk of the discussion was taken up by a very pointed demonstration which was reminiscent of Wolseley's abortive attempt to impress the Ashantis nearly twenty years before. But this time the Africans could hardly fail to get the point.

Suggestion for a Gatling-equipped Camel Corps 1872

Rhodes had brought an escort along with him and amongst their equipment was a Maxim gun. He took Sicgau into a field of mealies and there the gun was set up. It was then fired for a few seconds and great swathes of mealies were mown down. Then Rhodes turned to the chief and said: 'This is what will happen to you and your tribe if you give us any further trouble.'[14]

Sicgau was suitably impressed, but in that same year, in the area of Africa to which Rhodes gave his name, and unfortunately his ethos, his men had occasion to use their Maxims on men rather than mealies. In Rhodesia a small detachment of the British South African Police, using machine guns purchased by the British South Africa Company, were confronted by the Matabele led by their chief Lobengula. As early as April 1890 the assistant commissioner in the Bechuanaland Protectorate, aware of the realities of colonial control, had been writing home to anxiously ask, 'When may I expect the Maxim gun . . .?' By 1893 there were five of these guns available. In three battles, at Shangani on 24 October, Imbembesi on 1 November, and Empandine on 2 November, they were used to ruthlessly crush Matabele resistance. In a report on the second battle, Lieutenant-Colonel Willoughby fully acknowledged the part played by the machine guns: 'I think it is doubtful whether the rifle fire brought to bear would have succeeded in repelling the attack; the Matabele themselves have since stated that they did not fear our rifles so much, but that they could not stand against the Maxims.'[15]

The tragedy of the situation is revealed only too clearly in a contemporary account which appeared in the *Daily News*:

> Most of the Matabele had probably never seen a machine gun in their lives . . . Their trust was in their spears, for . . . they had never known an enemy able to withstand them. Even when they found their mistake, they had the heroism to regard it as only a momentary error in their calculations. They retired in perfect order and reformed for a second rush . . . Once more the Maxims swept them down in the dense masses of their concentration . . . It seems incredible that they should have mustered for still another attack, yet this actually happened . . . They came as men foredoomed to failure, and those who were left of them went back to a mere rabble rout.[16]

The Matabele themselves gave a moving account of their helplessness in the face of the machine guns. In it they pointed out that the whole war had started because the

British South Africa Company had chosen to interfere in an inter-tribal dispute:

> The Mashonas were Lobengula's subjects and the white man had no business with the Mashonas, to protect them or shield them from the King's justice. So *impis* had to be sent to punish these Mashonas, and they had collided with the white man. And the white man come again with his guns that spat bullets as the heavens sometimes spit hail, and who were the naked Matabele to stand up against these guns?[17]

Hutchison has recorded a macabre footnote to this slaughter of the bewildered Matabele. The Maxims were mounted on light artillery carriages and the man firing sat on a small saddle positioned at the end of the trail piece. Thus, 'after the Matabele War . . . there is something more than a legend which records that the Central African native fled at the mere rumour of the approach of a man who, taking up the common native posture for the relief of nature, could, in place of fluid, eject a death-dealing stream of metal.'[18]

One of the chief participants in this campaign had been Dr.Leander Starr Jameson. He had been so impressed with the performance of the Maxims that he attributed to them almost magical powers of protection. On 1 January 1896, when he rode forth across the Transvaal to attempt to bring armed support to the Uitlanders, the prosperous but disenfranchised gold seekers of Johannesburg, he took with him eight Maxims. Before the raid started, in an address to those who doubted his chances of success, he said of the guns: 'You do not know the Maxim gun. I shall draw a zone of lead a mile each side of my column and no Boer will be able to live in it.'[19] Unfortunately he neglected to take along adequate supplies of water, with which to cool the guns, and when they were brought into action they jammed almost immediately. His force was surrounded by Boer commandos under General Cronje, and Jameson was forced to surrender. Here one sees a rare example of an *over-estimation* of the machine gun's capabilities. And a fatal over-estimation it was too. Not only did it doom the raid to failure, but it also wrecked the political career of Jameson's friend, Rhodes, and brought a little nearer the prospect of war between the Boers and the English.

For the rest of the century, and into the twentieth, the use of machine guns was limited to consolidating the Europeans' hold on the African continent, and to the suppression of any native dissent. In 1897 Sir Arthur Hardinge gave a succinct definition of the true nature of European rule. His remarks referred to British policy in Kenya but they are applicable to all nations who had established a foothold on the 'dark continent'. As he said: 'These people must learn submission by bullets – it is the only school . . . In Africa to have peace you must first teach obedience and the only person who teaches the lesson properly is the sword.'[20] The choice of weapon is a little confused but the message is clear. The only adequate response to native discontent was violence. And despite Harding's rhetorical reference to the *arme blanche*, it was the machine gun that offered the most economical solution to the problem of keeping down the whole population of a continent with small bodies of police and soldiers.

They were used again, for example, in Rhodesia against the Matabele. In 1896, though hardly recovered from the massacres three years earlier, they rose up again. The rebellion took place in the immediate aftermath of the Jameson Raid and can be partly attributed to the fact that the whole area had been temporarily stripped of Europeans. But it seems likely that the abject fear of Jameson and his Maxims was also a contributory factor. For many of the Matabeles went to war this time sustained by the belief that the guns of the Europeans could do them no harm. At the trial of some of rebels it was stated by one Matabele witness that: 'The accused told them that when the white men crossed the river their bullets would turn to water and the Maxim could not fire any longer as there were no bullets left.'[21] Such beliefs are very reminiscent of those of Mwanga's men in Uganda, six years earlier. In each case the rumours were without much foundation, but it is interesting to note the speed with which the Africans correctly identified the central importance of the machine gun to the maintenance of European

Exactly the same kind of rumours were prevalent in German East Africa, what is now Tanzania, in 1905. The German Government had taken over control from the German East Africa Company in 1891 but the situation in the colony had not improved much. Eventually the natives were goaded into rebellion, on an unprecedentedly widespread scale. At least three tribes took part, the Mbunga, the Pogoro and the Ngoni, and certain scholars have discerned in this the first awakening of any kind of supra-tribal consciousness in the region. The ideological framework of the rebellion was a set of beliefs that came to be known as the Maji-Maji cult. The central belief was that no harm could come to the rebels because when they attacked, the white man's bullets would turn into water. But once again they were cruelly deceived, though these beliefs were strong enough to carry them through to the bitter end. On August 30, for example, '8,000 of the Mbunga and Pogoro tribes, armed only with spears, tried to assault the Mehenge fort, to drag away the machine guns with their bare hands.'[22] Of the Ngoni resistance, Count von Gotzen, the Governor of German East Africa, said charitably: 'The natives fought amazingly well, only retreating in the face of machine gun fire . . .'[23] The rebellion failed bloodily and the Africans learned to their cost that magical talismans alone could not alter the balance of military force, however perceptively it had been assessed. The only effective response to the machine gun and the general European preponderance of firepower was the resort to guerrilla warfare, and certain tribes did adopt this mode of combat.[24]

But it was not only the African who openly acknowledged the importance of automatic fire in the subjugation of the African colonies. The following letter to Cecil Rhodes from Sir Harry Johnston, an employee of the British South Africa Company, is chilling in its matter-of-fact assumption of the central role of violence in the day-to-day management of a particular region of Africa: 'One day I am working out a survey which has to be of scrupulous accuracy, and another day I am doing what a few years ago I never thought I should

The Sultan of Zanzibar with a Lewis Gun 1917

93

be called upon to do – undertaking the whole responsibility of directing military operations. I have even had myself taught to fire Maxim guns . . . I who detest loud noises . . .'[25] Another group who were prepared to come out into the open about the harsh realities of imperialism were certain British poets of the turn of the century. Perhaps because it was 'only poetry' people felt that their words had less real significance. Nevertheless, the message sometimes came through loud and clear. Even Rudyard Kipling, usually so piously smug about the duties of Empire and the thankless self-sacrifice involved in them, gave at least one scathing definition of Christian civilisation. In 1897 he wrote a poem called 'Pharaoh and the Sergeant', dedicated to the Sergeant-Instructors sent to Egypt to help train that country's ramshackle army. It begins thus:

> Said England unto Pharaoh, 'I must make a man of you,
> That will stand upon his feet and play the game;
> That will Maxim his oppressor as a Christian ought to do.'
> And she sent old Pharaoh Sergeant Whatsisname.

Halaire Belloc was equally blunt in a poem called 'The Modern Traveller'. In it a typical, somewhat languid colonial figure utters the perfect motto for the triumph of British imperialism:

> I shall never forget the way
> That Blood stood upon this awful day
> Preserved us all from death.
> He stood upon a little mound
> Cast his lethargic eyes around,
> And said beneath his breath:
> 'Whatever happens, we have got
> The Maxim Gun, and they have not.'

So the slaughter continued. In 1900 the Ashanti had once again to bear the brunt of British displeasure. At first the British fared rather badly. One force was besieged by the Ashanti in the fort at Kumasi, and the relief column, under Captain Aplin, which was sent from Lagos, was continually harassed by the natives. The main reason for Aplin's plight was that the inevitable Maxims accompanying the column were of an old and unreliable model, and whenever they were brought into action they never failed to overheat and jam. But Aplin's force eventually reached the fort and found that the plight of its occupants was not as desperate as had at first been feared. Yet again Maxim guns had saved the day, for 'at first the Ashantis had tried to attack the fort itself, but the machine guns on the bastions had proved too effective for them and they had settled down to a long and patient siege.'[26] More troops were thrown into the campaign as swiftly as possible, mainly from the Gold Coast Constabulary and the West African Frontier Force, all these units possessing large numbers of Maxims. They were used to great effect in the final battle of Aboasu in which Ashanti resistance was crushed for good.

As was common practice at this time, almost all the troops used in this campaign were Africans, only the officers being British. One of the most important duties of these officers was to operate the Maxim guns. It would clearly be too dangerous to teach natives, even though they might be wearing a British uniform, the secrets of the white man's ultimate weapon. But during one of Aplin's first attempts to relieve Kumasi there occurred one of the few instances of an African ever getting to actually fire a machine gun. At the height of the engagement all the Europeans had been killed or incapacitated and a native NCO took the chance to step into the breach. He managed to get the gun working and kept firing until it finally jammed. 'The history of this particular NCO is an interesting one; having fought the British in the Sudan he had conceived a profound respect for the Maxim gun and had walked across Africa and enlisted in the West African Frontier Force with the express purpose of working one.'[27] Once again one sees the central place of the machine gun in the Africans' analysis of the reasons for their conquest and subjugation.

But few other Africans ever got the chance to emulate his example. The Europeans jealously guarded both the machine guns themselves and the secrets of their operation. When the Dervish camp at Omdurman was overrun, amongst the vast amounts of military material found there were a few Gatlings and Nordenfelt guns, doubtless captured after the destruction of Hicks Pasha's army. But the Dervishes clearly had no idea how to fire them and they had remained uselessly in the rear. Similarly, in 1894, when the capital of the Itsekiri, at Brohemie in Nigeria, was captured by the British they found a machine gun rusting away in the huge armoury. Also in Nigeria, in 1906, the people of the Satiru tribe captured a Maxim when they wiped out a company of the West African Frontier Force, but the water-jacket had been slashed and the gun was unworkable. About the only occasion when machine guns were actually used against Europeans was in Uganda, in 1897, when Lugard's trusted Sudanese mercenaries mutinied and managed to lay their hands on one of the Maxim guns. But they did not use it very intelligently and staked their fortunes on a massed charge against the European position. They were quickly mown down by the defenders' own Maxims. Finally, on two occasions, machine guns were used in internecine disputes between the Africans themselves. The Ijesha of the Yoruba had a Gatling gun with them during the war with the Ibadan at Kariji, and another Gatling was used by the Amakari faction in New Calabar during the Civil War of 1879–82.

Lugard figured prominently in yet another campaign against the Africans. The setting was Hausaland (Northern Nigeria), in 1902–3, and the unfortunate natives were the Fulanis.[28] A first expedition was sent out under General Kemball who had with him a force of 1,000 men and six Maxims. But he did not fare particularly well, and in the following year another force, 800 men and five Maxims, was sent out under the command of General Moreland. This latter figure clearly set much store by the machine guns and felt, as Wolseley and Rhodes had done previously, that a demonstration of their power could do much to cow the Fulanis into submission. Thus when he had captured Kano, the Fulani Emir's capital, and found it necessary to discipline his men for looting, he chose a particularly macabre way of executing one of the principal offenders. An officer accompanying Morland's force said that: 'The execution was performed by half a belt being fired through a Maxim, in which case the inhabitants of Kano must have been most impressed not only with the efficiency but also with the extravagance of spending 125 rounds on the execution of a single malefactor.' But, impressed or not, the Fulanis insisted on fighting on and a final battle was fought at Sokoto on 15 March. The battle followed the classic course as these two eye-witness reports show. One of the Europeans said: 'As we approached close to the city hordes of horsemen and footmen armed with swords, spears, old guns and bows and arrows appeared, charging the square over and over again, only to be mown down by machine gun and carbine fire.' A Fulani participant, Hassan Keffi, put it even more simply: 'It was Sunday when they came. The guns fired "bap-bap-bap" and many hundreds were killed.'[29] Another Fulani expressed his bewilderment at this new type of warfare in which the two sides never even came to grips: 'War now be no war. I savvy Maxim-gun kill Fulani five hundred yards, eight hundred yards far away . . . It no be blackman . . . fight, it be white man oneside war. It no good . . . Previous slave-raiding not so bad as big battle where white man kill black man long way away. Black man not get come near kill white man. If he come near he die.'[30]

When the slaughter was over the victors could afford to be rather charitable to the survivors. As two historians have observed recently: 'After the Gatlings (sic) had dissolved the Fulani army . . . for the loss of one British officer from an arrow wound, the British told the surviving emirs that they had made a "plucky stand" – which was said to have consoled them considerably.'[31] It is interesting to note, as yet another indication of the central role of the machine gun in the colonisation of Africa, that the Maxims used in this campaign were carried on the heads of specially enlisted and trained bearers, and Vickers, Sons and Maxim had specially designed them so that they could be dismantled for this purpose.

As a final testimony to the ubiquity of the Maxim gun it is worth looking at the career of just one typical colonial officer at the turn of the century. Take the exploits of the Englishman, Seymour Vandeleur, whose *curriculum vitae* reads like an advertising brochure for machine gun manufacturers. In January 1895 he took part in an expedition to make reconnaissance across Lake Albert and down the Nile. A Maxim gun accompanied the eighteen members of the party. In February he was a member of a punitive force dispatched into Kabrega, in Uganda, and they took along with them several Maxim gun detachments. In April an even stronger force was sent out which included two Hotchkiss guns and three Maxims. In 1896 the Nandi incurred British displeasure and another expedition was mounted against them. And so was the inevitable Maxim. At one stage the Nandi attempted a frontal assault:

> Wheeling to the left by a common impulse, on they came, in spite of the Maxim, and charged down with great dash on our force which closed up to face the attack. It was a critical moment but . . . as the mass of natives approached our heavy fire began to tell . . . At last, wavering before the leaden hail which they had never before experienced, their ranks broke and they scattered in all directions, leaving many of their number on the ground.[32]

In the next year Vandeleur led an expedition against the Emir of Nupa, taking along seven companies of infantry, each with a Maxim gun attached, and one against the Belogun tribe of Ilorin, which consisted of 340 men and four Maxims. In 1898 he took part in the operations in the Sudan which culminated in the Battle of Omdurman. In 1899 he, some 2,000 others and six Maxims conducted mopping-up operations in Kordofan. Finally, in 1900, he took part in the

Boer War, joining the 1st Mounted Infantry Brigade, one of the few British units to include a reasonable proportion of machine guns (6,380 men and twenty Maxims).

Africa was not the only continent to experience the aggressiveness of British imperialism, but it does seem to have been only there that conditions were appropriate for the large-scale deployment of automatic weapons. There seems to be little mention of their being used to any extent in other parts of the world. In 1894 the Navy sent a four-gun Gardner battery to Burma after the occupation of Mandalay, for use against the Dacoits. The guns were carried on mules and could be assembled for action within thirty seconds. Even so the terrain was inimical to their use and they rarely saw action. Terrain was also the deciding factor with regard to tactics in India, particularly on the North-West Frontier. In these mountainous regions, howitzers, or screw guns, were invariably used in preference to machine guns. Only in Chitral did they make any impact at all. In 1895 a British relief column sent there found on a Pathan body a letter from a Scottish firm in Bombay which offered Umra Khan, the local chieftain, a wide range of arms and ammunition, including Maxim guns for a mere 3,700 rupees each. He never got the chance to take the offer up, largely because this was one of the few occasions on which the British in India themselves used machine guns.[33]

Five years later machine guns made a brief appearance in operations against the Chinese. During the Boxer Rising a heterogeneous assortment of soldiers and diplomats in Peking found themselves besieged by the rebels in the legation area. Their defence was much facilitated by the presence of an Austrian Maxim, an American Browning and a British Nordenfelt. The expedition which finally relieved them also made great use of the ten machine guns at its disposal.[34]

But perhaps the bloodiest excercise of the machine gun's potential in Asia occurred in 1904, during a British punitive mission to Tibet. Amongst the troops present was a detachment of the 1st Battalion of the Norfolk Regiment, and their equipment included two Maxims. In the final battle of the campaign between six and seven hundred Tibetans were killed, as against a mere handful of the British troops. At the height of the engagement:

> Out on the plain . . . the Maxims chattered vindictively. Under such fearful punishment no troops in the world could have stood their ground. It was not a battle but a massacre. Hadow, manning one of the Norfolk's Maxims, wrote to his mother that night: 'I got so sick of the slaughter that I ceased to fire, though the General's

order was to make as big a bag as possible.'[35]

The Tibetans' reaction to their impotence in the face of automatic fire made it even more difficult to simply keep on killing them, unless of course one shared the General's outlook and regarded them as so much game. For when they realised that they had no chance of actually reaching the British positions most of the Tibetans who were still alive simply bowed their heads, turned their backs on the soldiers and slowly walked away. In the space of a few traumatic seconds they had been taught a lesson about firepower that the British regular army had still not learnt in fifty years, and was not to learn until a little over ten years later.

In 1898 the US Government also took on colonial responsibilities of a sort when they annexed the Philippines, which had been under Spanish dominion for the preceding four centuries. In the following year they received the honour of having one of Kipling's poems addressed to them. In it he warned them of their heavy responsibilities:

Take up the White Man's burden –
Send forth the best ye breed –
Go bind your sons to exile
To serve your captives' need;
To wait in heavy harness
On fluttered folk and wild –
Your new-caught, sullen peoples,
Half devil and half child.

At the time few people in either Britain or the United States objected to such smug assumptions about the superiority of Western civilisation. Theodore Roosevelt described it as 'rather poor poetry, but good sense from the expansionist view point.' But there was in England a small group of radical anti-imperialists, known as the Little Englanders, who did find such patronising guff rather offensive. Their most prominent spokesman was Henry Labouchere, who as well as being an emphatic opponent of imperialism in general, was also keenly aware of the intimate links between 'paternal' British control and the power of automatic weapons. In a reply to Kipling's poem, published in *Truth*, a journal of which he was the editor, Labouchere banged his point home:

Pile on the Brown Man's burden!
And if ye rouse his hate,
Meet his old-fashioned reasons
With Maxims – up to date,

With shells and Dum-Dum bullets
A hundred times make plain
The Brown Man's loss must never
Imply the White Man's gain.

In the same journal he also lampooned writing that was generally regarded as being even more sacred than Kipling's verse. His *Pioneers' Hymn* began thus:

Onward Christian Soldiers, on to heathen lands,
Prayer-books in your pockets, rifles in your hands,
Take the glorious tidings where trade can be done:
Spread the peaceful gospel – with a Maxim gun.

Labouchere was also a Member of Parliament and in a debate of 1893 he again attacked those who condoned extreme brutality in the name of the spreading of the Christian message: 'During the occupation of Uganda, the cause of Christianity has suffered, and to my mind that is not surprising when Christianity is associated in mens' minds with Maxim guns.' In 1894, in another debate in the House of Commons, Sir Charles Dilke, an ally of Labouchere on this point, made an even more telling comment: 'We are everywhere bound to follow missions with our arms and flag . . . The only person who has up to the present time benefited from our enterprise in the heart of Africa has been Mr.Hiram Maxim.' One of our nominal aims, it seemed, had been the propagation of the Gospel, but what, he wondered, would have been England's fate 'if Augustine had landed in Kent with Maxim guns?'[36] In January of the same year the *Westminster Budget* also took up the struggle. They printed a caricature of Jameson riding out to war with the initials BSAC stamped on his horse's harness. Underneath was a caption which read:

Through shot and hell to glory – and Bulawayo!'
To the *Lancet*; 'In my humble opinion Maxim bullets
Are a decidedly effective remedy for Matabele evils.'[37]

Such propaganda did have some effect in arousing public opinion, and as we advanced into the new century colonial blood-letting gradually diminished. But this was as much as anything attributable to the fact that the previous years of ruthlessness had temporarily beaten many tribes into submission. Only when the Europeans' own parcelling out of the continent began to foster supra-tribal feelings of national identity could African resistance find a strong enough base to flare forth once more. Until that time the machine guns were always kept well-oiled. Even in 1906 one of the last stands of the Zulu nation was savagely put down with artillery and machine guns. At the Battle of Mome Gorge the order was given that prisoners should not be taken, and the whole force of Zulus, surrounded and trapped by their adversaries, were mown down. When hearing of the massacre, John X. Merriman, a liberal South African politician, was prompted to echo the old complaints of the Little Englanders: 'We have had a terrible business in Natal with the natives. I suppose the whole truth will never be known, but enough comes out to make us see how thin the crust is that keeps our Christian civilisation from old-fashioned savagery – machine guns and modern rifles against knobsticks and assegais are heavy odds and do not add much to the glory of the superior races.'[38]

By now, then, the picture should be clear. The machine gun was a vitally useful tool in the colonisation of Africa. Time and time again automatic fire enabled small groups of settlers or soldiers to stamp out any indigenous resistance to their activities and to extend their writ over vast areas of the African continent. Yet, as has already been indicated, this gruesome test of the machine gun's efficiency had almost no effect upon the military High Command in Britain. For the 'brass' back home automatic fire was still a tiresome gimmick, not likely to play any significant role upon a European battlefield. The reasons for this complacency are not hard to find.

They are to be found in the ideology of British imperialism, whose very essence was an unquestioning belief in the innate superiority of the white race, and the British in particular. Without such beliefs it would have been impossible for the orginal colonisers to set such a low price on African lives. For only by holding them so cheap could the slaughter of the natives seem to be morally acceptable. The belief in white supremacy was the very bedrock of imperialist attitudes, and is evident in all their manifestations. At best the Europeans regarded those they slaughtered with little more than amused contempt. Thus Winston Churchill in a letter home from the Sudan:

> It is like a pantomime scene at Drury Lane. These extraordinary foreign figures . . . march one by one from the dark wings of Barbarism up to the bright footlights of civilisation . . . And the world audience clap their hands, amused yet impatient . . . and their conquerors, taking their possessions, forget even their names. Nor will history record such trash . . . Perhaps the time will come when the supply will be exhausted

101

and there will be no more royal freaks to con-
quer . . . The good old times will have passed away,
and the most cynical philosopher will be forced to
admit that though the world may not be much more
prosperous it can scarcely be so merry.[39]

This feeling of contempt is obviously functional within a
colonial situation. It dehumanises one's opponent and
makes it easier to seize his lands and his labour, and to stamp
ruthlessly on any sparks of resistance. Thus when it becomes
necessary to kill those who stand in one's way, the problem is
seen in technical rather than human terms. It is simply a
matter of 'bagging' as many natives as possible with the
minimum effort. The machine gun filled these requirements
admirably.

Its widespread adoption was further facilitated by the fact
that many of the colonial units were not made up of regular
soldiers. They were ordinary civilians who only grabbed a
weapon when the need arose. They were free of the centuries
of tradition that so blinkered the regular establishments in
Europe and were able to assess automatic weapons in terms
of a crude cost-benefit analysis. Such thinking was not without effect on the regular soldiers who were sent overseas.
Because they regarded the Africans as weird eccentrics,
hardly even human beings, they could look on colonial warfare as an amusing diversion that had little in common with
the 'real' wars that had been fought in Europe and might
have to be fought in the future. Thus, because the machine
gun had become so much a part of these imperialist sideshows, it came to be regarded, by definition, as a weapon
that had no place upon the conventional battlefield. The
European was obviously superior to the African, so why
would he ever be so stupid as to be baulked by a weapon that
was really only good for bowling over 'niggers' and 'Kaffirs'?
Of all the chickens that came home to roost and cackle over
the dead on the battlefields of the First World War, none was
more raucous than the racialism that had somehow assumed
that the white man would be invulnerable to those same
weapons that had slaughtered natives in their thousands.

So the machine gun came to be regarded as a weapon
suitable only for use against African natives and the like. Of
the Ashanti campaign of 1873, the *Army and Navy Journal* said:
'We are not surprised that the Ashantees were awestruck
before the power of the Gatling gun. It is easy to understand
that it is a weapon which is specially adapted to terrify a
barbarous or semi-civilised foe.' In fact, as has already been
mentioned, the Ashanti were far from awestruck by this
particular demonstration of automatic fire, but the essential

point still remains. In 1879 the *Army and Navy Gazette* applauded the fact that Gatlings were to accompany the expedition to Zululand: 'Gatlings are the best of all engines of war to deal with the rush of a dense crowd.' In a lecture to the Royal United Services Institution in 1885 Chelmsford looked back on this campaign and observed that: 'Machine guns are, I consider, most valuable weapons for expeditions such as that which we had to undertake in Zululand, where the odds against us must necessarily be great . . . If a machine gun can be invented that may safely be entrusted to infantry soldiers to work, and could be fired very much as one grinds an organ, I am satisfied of its great value.'[40]

In his autobiography Hiram Maxim himself revealed that his own gun had been designed with such a role very much in mind. He tells of an early discussion with Wolseley, shortly after the first working model had been demonstrated:

> Lord Wolseley was one of the cleverest and brightest military men that I have ever met. I often met his lordship, and on one occasion he commenced to discuss the machine gun question. If machine guns were to be used in the service he saw no reason why they should not have a longer cartridge and a longer range than the infantry rifles . . . I told him that such a gun would not be so effective as the smaller gun in stopping the mad rush of savages, because it would not fire so many rounds in a minute, and that there was no necessity to have anything larger than the service cartidge to kill a man.[41]

This attitude did not only prevail in England. Though the Germans began to realise the machine gun's value before the outbreak of the First World War, they too had been blinded by racialist assumptions. As a deputy in the Reichstag said in 1908: 'A year ago people in military circles were not so conscious of the value of machine guns as they are today. Then there were many people, even in the German Army, who still regarded the machine gun as a weapon for use against Hereros and Hottentots.'[42] In the British Army such beliefs persisted, though a few warning voices were raised. In 1900 Spencer Wilkinson modified his earlier views and observed sadly that: 'The army ought to have learnt from its own experience in the Sudan that the bravest and most athletic troops cannot possibly, however fleet and sound-winded, carry the knife or spear within the reach of a line of riflemen.'[43] Not to mention machine gunners. But such warnings went unheeded. How could anyone have the temerity to suggest that the British soldier might have something to learn from the fate of a mob of Dervishes?[44]

But there was another reason for the general desire to ignore the evidence of the machine guns deadly potential. In England the vast proportion of the ruling élite, government, bureaucracy and armed services had been educated in the public schools. This connection was particularly evident in the armed services. An examination of the future careers of boys leaving Harrow and Rugby between 1830 and 1880 shows this clearly. In 1830, 12 per cent of the leavers joined the armed forces; in 1840, 17 per cent; in 1850, 27 per cent; in 1860, 21 per cent; and in 1880, 14 per cent.[45] Moreover, of the 504 cases represented here, only one was not an officer, and 65 of the rest reached the rank of full colonel or its naval equivalent. The situation was much the same in the newer public schools that sprang up over the century. Of a total of 4,227 boys leaving Cheltenham between 1841 and 1910, 2,896 joined the armed forces, and the figures for Clifton School were almost identical.[46] Another writer has pointed out that, in 1891, 55 well-known public schools or universities supplied the total 373 cadets who entered Sandhurst that year.[47]

One of the most interesting aspects of the public schools was their emphasis upon games and the philosophy of playing hard but playing fair that went with them. Little research has been done on this topic at present and it is difficult to formulate more than the most tentative conclusions. But it does seem clear that for many Englishmen at this time there was a very real feeling that the principles of the game were those most applicable to the organisation of one's life. Even in 1852 General Sir John Burgoyne posed his objections to the introduction of compulsory education tests in the army in this way: 'Such tests are uncalled for, delusive, mischievous . . . At the public school will be found one set of boys who apply to their studies, and make the greater progress in them; another set take to cricket, boating, swimming etc. Now of the two I should prefer the latter, as much more likely to make good officers . . . Other matters are of no absolute use to them in their profession.'[48] The connection between life and sport was emphasised more and more as the years passed. 'From 1870 or thereabouts, at a conservative estimate, there was a subtle but organised drive by authority to sublimate the boy's self to a team; and this way of life, which resembled nothing so much as a human anthill heaving for a common purpose, was elevated by its supporters into a major principle of education.'[49]

Baden-Powell, not unnaturally, expressed this philosophy as well as anyone. He had participated in the expedition against the Matabeles in 1896, and at one stage was prompted to observe:

> Your Englishman . . . is endowed by nature with a spirit of practical discipline . . . Whether it has been instilled into him by his public school training, by his football and his fagging, or whether it is inbred from previous generations of stern though kindly parents, one cannot say; but, at any rate, the goodly precepts of the game remain the best guides: 'Keep in your place', and 'Play not for yourself, but for your side.'[50]

The relationship between sport and warfare itself was seen by one Captain Guggisberg to be perfectly self-evident. In a book called, rather hopefully, *Modern Warfare or How Soldiers Fight*, written in 1903, he made the following analogy:

> An army tries to *work together* in battle . . . in much the same way as a football team *plays together* in a match; and you need scarcely be told what an important thing that is if you want to win. The army *fights* for the good of its country as the team *plays* for the honour of its school. Regiments *assist* each other as players do when they *shove together* or *pass the ball* from one to another; exceptionally gallant *charges* and heroic *defences* correspond to brilliant *runs* and fine *tackling*. All work together with one common impulse, given to the army by its general, to the team by its captain.[51]

The poetry of someone like Sir Henry Newbolt is also very relevant in this respect. Many of his poems lay great stress upon this link between an Englishman's duty and the 'goodly precepts of the game', as taught within the public schools. In particular he addressed his poems to those English soldiers fighting overseas for King and Empire. The publisher's advertisement to the collection *Clifton Chapel* (1908) makes this quite plain: 'The whole collection deals with English school life, mainly in its Imperial aspect; it is published by special request for the users of Clifton College, and will, it is hoped, commend itself to members of other Public Schools.'[52] His immortal hymn to the Public School spirit, *Vitai Lampada* is probably the best known example of his work:

There's a breathless hush in the close tonight –
Ten to make and a match to win –
A bumping pitch and a blinding light,

An hour to play and the last man in.
And it's not for the sake of a ribboned coat,
Or the selfish hope of a season's fame,
But his captain's hand on his shoulder smote,
'Play up! Play up! and play the game!'

The sand of the desert is sodden red –
Red with the wreck of a square that broke –
The Gatling's jammed and the colonel dead,
And the regiment blind with dust and smoke.
The river of death has brimmed its banks,
And England's far and Honour a name,
But the voice of a schoolboy rallies the ranks,
'Play up! Play up! and play the game!'

Such a poem is a quintessential product of its time. No one but a late nineteenth-century Englishman could move so easily straight from a description of a cricket match to the predicament of broken troops in Africa. And who else could think that the same inane exhortation could hearten a hapless tail-ender and win a battle? Yet Newbolt is a little atypical in so far as he actually mentions the Gatling gun. For one by-product of this stress upon the cult of playing the game was the accompanying notion of 'fair play'. Though this notion was often not strong enough to resist the demands of imperial expediency, it did leave many Englishmen with a residual sense of unease about the atrocities that were being committed in Africa. For the widespread use of automatic weapons against adversaries armed only with clubs and spears could not by any stretch of the imagination be regarded as fair play. To a large extent, consciences could be calmed by the knowledge that Africans were not quite human, and therefore beyond the pale of Imperialist morality. Even so, consciences could be mollified yet more if they quietly ignored the fact that the breathtaking victories of the British were largely attributable to vastly superior firepower. In this respect it is perhaps significant that Newbolt does not let his public school archetype on the scene until the Gatling has jammed.

But an undue emphasis upon the role of firepower in these African campaigns not only offended Victorian consciences, it also tended to undermine faith in the British soldier as such. Most people at home still thought in terms of a European war and thus liked to believe that they had an army that was man for man superior to any other army on the continent. Of this period Corelli Barnett has written:

The campaigns of the three Victorian heroes,

Roberts, Wolseley and Kitchener, represented essentially all that British people knew about modern war . . . There was emphasis upon the man rather than the system . . . on minute casualties and easy victories . . . War against savages could not really test an army. The colonial triumphs created a dangerous impression at home that wars were distant and exotic adventure stories, cheaply won by the parade-ground discipline of the British line, that to win a modern war you called for a hero.[53]

This is true enough, and the effects of such a belief will clearly be seen in the next chapter. But the interpretation of Victorian successes did not merely *create* a myth, it was also a response to a *demand* for a myth. Britain's last experience of war in Europe had been the Crimean War, and people were eager for a new mythology of British invincibility that would wipe out the humiliation of that particular campaign. The wars in Africa certainly produced stunning victories in terms of casualty figures. But it was necessary to ignore the fact that machines rather than men had been responsible for these victories.

Notes

1. Clarke, op.cit., pp.63 and 78–9.
2. V.A.Majendie, *The Arms and Ammunition of the British Service*, Cassell, London, 1872, p.viii.
3. Wahl and Toppel, op.cit., p.67.
4. D.R.Morris, *The Washing of the Spears*, Jonathan Cape, London, 1966, pp.457 and 570.
5. Hutchison, op.cit., p.39.
6. Wahl and Toppel, op.cit., p.86.
7. Ibid., p.107.
8. Symons, op.cit., p.217.
9. Pridham, op.cit., p.42.
10. G.W.Steevens, *with Kitchener to Khartoum*, Blackwood and Sons, London, 1898, p.300.
11. Maxim, op.cit., p.258.
12. C.Miller, *The Lunatic Express*, Macdonald, London, 1972, pp.230–31.
13. J.Listowel, *The Making of Tanganyika*, Chatto and Windus, London, 1965, p.28.
14. J.G.Lockhart and C.M.Woodhouse, *Rhodes*, Hodder and Stoughton, London, 1963, p.192.
15. S.Glass, *The Matabele War*, Longmans, London, 1968, p.195.
16. Hutchison, op.cit., p.35.
17. T.Ranger, *Revolt in Southern Rhodesia, 1897–7*, Heinemann, London, 1967, p.121.

18. Hutchison, op.cit., p.44.
19. Lockhart and Woodhouse, op.cit., p.310.
20. S.Kiwanuka, *A History of Buganda*, Longmans, London, 1971, p.259.
21. Ranger, op.cit., p.176.
22. J.Iliffe, 'Tanzania under German Rule', in B.A.Ogot and J.A.Kieran (eds.), *Zamani, a Survey of East African History*, Longmans, London, 1968, p.295.
23. Listowel, op.cit., p.41.
24. Ellis, John, *Short History of Guerrilla Warfare*, Ian Allan, 1975.
25. R.A.Huttenback, 'G.A.Henty and the Vision of Empire', *Encounter*, July 1970, p.50.
26. A.Lloyd, *The Drums of Kumasi*, Longmans, London, 1965. p.181.
27. F.Myatt, *The Golden Stool*,William Kimber, London, 1966, p.91.
28. See M Crowder, *The Story of Nigeria*, Faber and Faber, London, 1966.
29. D.J.M.Muffett, *Concerning Brave Captains*, Andre Deutsch, London, 1964, pp.98, 132 and 133.
30. Brigadier-General F.P.Crozier, *The Men I Killed*, Jonathan Cape, London, 1937, p.149.
31. R.Lewis and Y.Foy, *The British in Africa*, Weidenfeld and Nicolson, London, 1971, p.183.
32. Col.F.I.Maxse (ed.), *Seymour Vandeleur: the Story of a British Officer*, National Review Office, London, 1905, p.64.
33. See Major-General J.G.Elliott, *The Frontier 1839–1947*, Cassell, London, 1968, and A.Swinson, *North-West Frontier*, Hutchinson, London, 1967.
34. See P.Flemming, *The Siege at Peking*, Rupert Hart-Davis, London, 1959.
35. P.Fleming, *Bayonets to Lhasa*, Rupert Hart-Davis, London, 1961, p.151.
36. Miller, op.cit., pp.269 and 279.
37. J.Selby, *Shaka's Heirs*, Allen and Unwin, London, 1971, p.219.
38. S.Marks, The Zulu Disturbances in Natal, in R.I.Rotberg (ed.), *Rebellion in Black Africa*, Oxford University Press, New York, 1971, p.59.
39. F.Woods (ed.), *Young Winston's Wars*, Sphere Books, London, 1972, pp.195–6.
40. Wahl and Toppel, op.cit., pp.69, 86 and 89.
41. Maxim, op.cit., p.182.
42. McCormick, op.cit., p.84.
43. Wilkinson, op.cit., p.409.
44. On this point see also J.Wheldon, *Machine Age Armies*, Abelard-Schuman, London, 1968, p.5.
45. T.W.Bamford, *The Rise of the Public Schools*, Nelson, London, 1967, p.210.
46. Ibid., p.219.
47. Razzell, op.cit., p.257.
48. Vagts,op.cit., p.172.

49. Bamford, op.cit., p.83.
50. J.Lord, *Duty, Honour and Empire*, Hutchinson, London, 1971, p.109.
51. Clarke, op.cit., p.132.
52. J.A.V.Chapple, *Documentary and Imaginative Literature 1880–1920*, Blandford Press, London, 1970, p.158.
53. C.Barnett, *Britain and her Army*, Allen Lane, London, 1971, p.324.

V The Trauma: 1914-18

'We were very surprised to
see them walking, we had
never   seen   that
before ... They went down
in their hundreds. You
didn't have to aim, we just
fired into them.'

A German machine gunner in 1916.

During the First World War the massive contradiction that
one has seen building up in the previous chapters was forced
to resolve itself. On the one hand, by 1914, the Europoean
powers were sufficiently advanced industrially and tech-
nologically to wage a war in which they could effectively
mobilise the full potential of the nation. Productive
capacities, administrative expertise, bureaucratic centrali-
sation, and methods of supply had all reached a point where
the nation as a whole could be involved in some aspect of the
struggle. The kind of conditions that have already been
discerned in America during the Civil War, and which were
shown to have contributed to that struggle's unpre-
cedented scale and bloodiness, were even more apparent in
the Europe of 1914. No matter what mens' preconceptions
about a limited European war, generally expected to be over
in a few months, the sheer potential of the combatants, in

terms of material and manpower, dictated a prolonged, savage struggle. Much is made of the distinction between limited wars and total wars. Those of the eighteenth century, or the guerrilla wars of today are usually put in the first category, the two world wars in the second. But, discounting the modern possibility of a war that would destroy civilisation itself, it is surely true that in most wars the adversaries fight to the limits of the weaker side. Thus for Frederick the Great the Seven Years War was a total war, as were the Napoleonic Wars for Bonaparte, or the Vietnamese

French machine-gun crew (La Mitrailleuse: C.R.W. Nevison 1915)

Wars for the Vietnamese. To this extent all wars are total wars.

In the First World War, because the countries involved on each side, or at least the major participants, were powerful industrial societies, each had to mobilise its full potential. The war became one of attrition in which victory could not be gained until one's enemy had been bled dry. His capacity to throw more and more into the field – the British began the war thinking 100,000 men would suffice; by September 1914 they had 500,000; by December 1915 the figure had reached three and a half million – meant that a premium would inevitably be placed on those weapons that could annihilate the enemy as cheaply and as quickly as possible. During the First World War the machine gun was the most important such weapon. Had a more effective way of controlling gas been found, it would no doubt have superseded the former weapon. During the Second World War a similar logic produced a thousand bomber raids and, eventually, the nuclear bomb. Basically then the increased reliance on the machine gun that was a feature of the 1914–18 War was an inevitable consequence of the necessity to wear one's enemy down as cheaply and completely as was then technically possible. By the end of the war every side was producing machine guns in unprecedentedly huge quantities.

The Emergence of the Machine Gun

The French had only adopted the machine gun in 1910, at which time the General in charge of Infantry remarked: 'This device makes no difference at all.'[1] In 1914 the whole French Army could muster no more than 2,500 of these weapons. Yet in 1918, as has been shown already, the massive output of the Hotchkiss factory had created an appreciable number of millionaires.

The British began the war with a mere two guns per battalion. But on 22 October 1915 the order for the setting up of a separate Machine Gun Corps was issued. Over the next months the heavy machine guns of the various divisions were redesignated as companies in the new Corps. By November 1918 it had a complement of 6,432 officers and 124,920 others. At the same time, to replace the absent Vickers guns, the battalions began to be equipped with Lewis light machine guns. In late 1915 there had been four of these guns in each battalion. In June 1916 this number was increased to eight, and in July a battalion headquarters section of four more guns was added. In December there was one Lewis gun for every four platoons, and by 1918 one for every two, as well as four others specifically allocated to anti-aircraft duties. Even the cavalry were forced to give way to the demand for more automatic weapons. In 1918 the Life Guards and the Blues were turned into lorry-borne machine gunners. They were formed into battalions of 64 guns each and were used by GHQ as a general reserve.

Thanks to the efforts of the men of the Machine Gun Corps and various officers in the line at battalion and divisional level, machine gun tactics also became more sophisticated. The Corps staff worked out complicated patterns of mutually supporting fire, so that the Vickers were not only used to repulse enemy assaults, but could also keep up a continual harassing fire on his trenches and rear areas. Lewis guns too were used with more imagination by some units. In July 1917 General Jack's men had already made them an integral part of the attack. Speaking of his brigade's assault training at this time, Jack wrote: 'Since I think it imperative to stop Germans who escape the artillery from coming into action at point-blank range, we are heavily manning our leading lines with Lewis guns. Each gun is carried by two men, the first with arm looped round the muzzle to steady it, the second holding the butt to fire.[2] But

A very anachronistic impression of a machine-gun action in August 1914

such sophistication was a little premature for most units, and it was not until 1918, and the larger allocation of Lewis guns, that the army in general began to incorporate the light machine gun into its infantry tactics.

In fact it was the Germans who led in this respect. Under the guidance of General Huitier, serving in Russia, they developed so-called 'infiltration tactics'. An attack was preceded only by a very short bombardment, and the leading infantry, usually attacking at night, were made up of small groups of 'storm troopers', whose main equipment was composed of light machine guns or automatic rifles. Such units were told to push forward as rapidly as possible, and to leave the reduction of any strongpoints to the infantry behind. These tactics were a great success in the last German offensive of 1918, but the army as a whole simply did not have the necessary reserves of men or even determination to carry the thing through. And by this time the Americans had entered the war.

The Germans also, like everyone else, found themselves obliged to greatly increase the numbers of machine guns as the war progressed. In earlier chapters I have had cause to mention the greater German awareness of the potential of automatic weapons, as compared to the attitude of the British. Yet, in the last analysis, the difference was only marginal. In terms of theory, the Germans were undoubtedly somewhat more advanced than the Allies, yet at the beginning of the war they too found that their complement of machine guns was woefully inadequate to the demands of modern warfare. It has often been asserted that the Germans entered the First World War with a higher proportion of machine guns than the Allies. In fact, 'contrary to the general British view, the Germans began the war with roughly

A Hotchkiss gun being demonstrated to the British

the same proportion of machine guns to their infantry, but they had organised their machine gunners in companies, which, by acting together, often gave the impression of a larger total number of guns per thousand men.'[3] They soon began to increase this number. Even in 1914 there were also eleven separate machine gun companies attached to the cavalry divisions, which were moved around independently to cover weak points in the line. By 1916 each division contained 72 heavy guns, and by 1918 the number had reached 350. Also in 1916 the independent companies were grouped together in units of three, known as *Maschinengewehr-Scarfschützen-Abteilungen*, and by 1918 there were 87 such units. The Germans also made much use of light machine guns and automatic rifles. In 1915 whole battalions were formed, equipped with 129 Muskete automatic rifles. In the summer of 1916 there were also formed light machine gun sections armed with the Bergmann gun. Each section had nine guns and there were eventually 111 such sections in existence.

On the British side in particular these increases seem to have been pushed through by the civilians, and a few enlightened military thinkers, rather than by the High Command itself. For, despite massive industrial and technological advances, despite the manifest efficacy of automatic fire in Africa, the armies of 1914 were still obsessed with past traditions and obsolete conceptions of warfare. The officer corps still looked for battlefields where there was a place for individual acts of heroism. The advent of industrialised society had in fact meant the end of the type of war such men dreamed of. Battle was now simply a matter of numbers, and the heroism of the individual, or even the unit, was now an irrelevancy. But old habits die hard. Even the British experiences in the Boer War did not make much of a dent in the complacency of the High Command. At the time there was much concern. A Commission of Enquiry was set up and people claimed that we had been taught a valuable lesson. Kipling was perceptive on this point:

Let us admit it fairly, as a business people should,
We have had no end of a lesson; it will do us no end of
    good.
It was our fault and our very great fault, and not the
    judgement of Heaven.
We made an army in our own image, on an island nine
    by seven,
Which faithfully mirrored its makers' ideals, equip-
    ment, and mental attitudes —
And so we got our lesson: and we ought to accept it with
    gratitude.

But though Kipling was right to point out that the army that had gone to South Africa was shaped by the attitudes of its leaders, he was wrong to suppose that these leaders were taught any lasting lesson. Despite the work of Lord Esher's War Office (Reconstitution) Committee, and the reforms pushed through by Lord Haldane, the British Army still remained pathetically small, by European standards, and, after an initial burst of enthusiasm, badly neglected. Referring to Haldane's successor, J.E.B.Seeley, the military writer, Repington, said that he was possessed of 'that attitude of bland assurance and complacent optimism which enables him . . . to place a gloss upon every defect, and to ignore the fact that since he assumed the reins of office nothing has been done – nothing of any kind – except to allow matters to go from bad to worse unchecked.'[4]

Essentially, then, things went on as before. The same kind of attitudes continued to prevail that had already done for a hundred years and more. For the average officer, and certainly the more senior ones, war was still a matter of will, in which the grit and resolution of the individual soldier counted for much more than any piece of machinery. Anything that was not compatible with this conception, anything that seemed to threaten the centrality of man upon the battlefield, was dismissed as being an unmilitary gimmick. Chief among these was the machine gun. One of the first reactions to automatic fire was given by an American journalist in 1863. Speaking of the Gatling gun he said: 'But soldiers do not fancy it. Even if it were not liable to derangement, it is so foreign to the old familiar action of battle – that sitting behind a steel blinder and turning a crank – that

Members of the Machine Gun Corps training

enthusiasm dies out; there is no play to the pulses; it does not seem like soldiers' work.'[5] These remarks sum up British Army reactions to the machine gun right through until the last months of the First World War.

Nor was it simply the machine gun that was in disrepute. The traditionalism of the army leadership made it suspicious of technology in general. Even the telephone, it seems, was rather *infra dig*. Of the planning of an artillery barrage in July 1916, during the Battle of the Somme, one learns that:

> No preliminary warning of any sort had been given; no opportunity made to permit the various commanders to go forward to make a reconnaissance while the artillery plan was being prepared. Indeed, neither of the brigade commanders was consulted in any particular about the fire plan though each had a telephone in use. But this was so often the custom of the time.[6]

Of this contempt for any kind of mechanical methods, Brigadier-General Baker-Carr, who was in charge of machine gun training during the first years of the war, has said: 'This feeling towards mechanical methods as opposed to hand skill may possibly go far to account for the widespread resistance which I encountered when striving to ensure the recognition of the machine gun as one of the most important weapons of the foot-soldier . . . and for the maddening want of vision on the part of those with whom decisions rested.'[7] It is perhaps unfair to suggest that this prejudice against technology was only to be found amongst prominent military men. Though it was undoubtedly of a much more extreme nature in these quarters, and lasted longer, it is nevertheless true that, with regard to war at least, the public at large was still very prone to bursts of pre-industrial romanticism. I.F.Clarke has shown that the man in the street at this time did show a lively interest in the nature of the next European war. But given all this speculation, it is rather ironic to note that by far the most enduring legend of the war years was the supposed appearance of the Angel of Mons. The legend grew out of a short story by Arthur Machen in the *Evening News*, which envisions a mode of warfare that even the British General Staff might have deemed a little anachronistic. At one stage in the battle, according to the narrator of the story:

> He heard, or seemed to hear, thousands shouting:
> 'St.George! St.George!'
> 'Ha! messire; ha! sweet Saint, grant us good deliverance!'
> St.George for merry England!'

Harow! Harow! Monseigneur St.George succour us.'
'Ha! St.George! Ha! St.George! a long bow and a strong bow.'
And as the soldier heard these voices he saw before him, beyond the trench, a long line of shapes, with a shining about them. They were like men who drew the bow, and with another shout, their cloud of arrows flew singing and tingling through the air towards the German hosts.[8]

Given the prevalence of such attitudes it is perhaps surprising that use was ever made of automatic weapons. Certainly its prejudice against them was not entirely overcome even after their battlefield dominance had become apparent. In September 1914 the Germans had reached the Aisne and turned to face the pursuing British and French. The Germans dug themselves in and set up machine gun posts. The Allies found out that they were unable to penetrate the defensive line and they in turn dug themselves in. The war of movement was over. Firepower totally dominated the battlefield and the opposing generals stared impotently at each other, not really understanding why their grandiose strategic theories had proved so misleading.

With regard to the machine gun this incomprehension manifested itself in two ways. Throughout the war many leading generals neither saw the need for machine guns in our own army, nor for a reappraisal of our tactical thinking because of their presence with the opposing forces. For the

A Vickers crew in action

first two or three weeks of the war there was perhaps some slight justification for the former attitude. It will be remembered that the musketry experts at Hythe had been forced to make the British rifleman into the most efficient in Europe. For a short time this expertise was sufficient. Indeed the Germans were deceived into thinking that the BEF had a bigger complement of machine guns than they had. One report from the first weeks told that: 'Over every hedge, bush, and fragment of wall floated a thin film of smoke betraying a machine gun rattling out bullets.' But this misconception could not last for long. The most effective riflemen in the world are simply no match for large numbers of machine guns, even indifferently handled. This potential gulf in firepower became even more marked as the BEF began to sustain heavy casualties and Kitchener's badly trained volunteers were rushed to France. For it was impossible, in a few weeks, to bring these men up to anything like the standard of the long-term regulars. Whereas a large proportion of the BEF had been able to place thirty shots within a target in one minute, it was found that very few of the new levies could even manage ten shots. Without machine guns British firepower was clearly not adequate for the type of war that was now developing. And even before the decimation of the old regulars, one might claim that the generals should have discerned some significance in the fact that the first two VCs to be won at the Battle of Mons were awarded to machine gunners.

But in the first years of the war they remained blind to the obvious necessity for a fast increase in the number of automatic weapons. Indeed, were it not for certain more perspicacious civilians in Britain, it seems doubtful whether the army would ever have had an anything like an adequate supply:

> Lloyd George . . . refused to accept the War Office estimates of future needs. He was equally contemptuous of generals in the field . . . He enquired how many machine guns were needed. Haig replied: 'The machine gun was a much over-rated weapon and two per battalion were more than sufficient.' Kitchener thought that four per battalion might be useful, 'above four may be counted a luxury.' Lloyd George told his assistants: 'Take Kitchener's maximum; square it, multiply that result by two – and when you are in sight of that, double it again for good measure.'[9]

Even though Kitchener was in favour of some slight increase in the allotment per battalion, he showed a typical military

120

vagueness about how they should be distributed. Sir Eric Geddes has left a description of the difficulties of communicating with these pre-technological soldiers:

> I told Kitchener that rifles and machine guns were the same as shillings and pounds: that nine rifles were equal to a Lewis automatic gun and thirteen rifles to a Vickers machine gun in the productive effort required for their manufacture. I wanted to know the proportions of each required for nine months ahead so that I could make my plans. His reply was: 'Do you think that I am God Almighty that I can tell you what is wanted nine months ahead? . . . I want as much of both as you can produce.'[10]

The generals were men to whom arithmetic, no matter how simple and obvious, meant absolutely nothing. In November 1915 Lloyd George pointed out in a memorandum to the War Committee that 50,000 machine gunners could easily do the work of at least a quarter of a million riflemen. General Sir Archibald Murray was of the opinion that the Adjutant-General would not be willing to spare the men necessary during the period of training at the Grantham Machine Gun School, preferring to send them to the front as simple riflemen. The War Committee was forced to order the Army Council to have 10,000 men continuously under instruction.

Abortive German Attack at the Second Battle Ypres 1915

In France itself, resistance to the machine gun was even more bitter. Baker-Carr has left behind a depressing record of his struggles with the High Command up to the end of 1915. He was originally put in charge of training the machine gunners of the Fifth Division alone, but found other divisional commanders so enthusiastic that he managed to extend his activities to cover the whole of the army on the Western Front. The divisional commanders urged him to double the number of men at the School, but Baker-Carr found it very difficult to lay his hands on an adequate number of machine guns. But eventually, 'even the General Staff could be made to understand that it was impossible to give instruction in a weapon unless the weapon itself was available.' But there was little beyond that that they could be made to understand. 'Already I was urging the advisability of doubling the number of machine guns per battalion i.e. from two to four. I had put forward the suggestion to GHQ and had been promptly told to mind my own business.'

But he pressed again and again and was eventually asked to submit a memorandum. This he did and GHQ promptly stuck it in their files. Again he pestered them for weeks and finally it was returned with Headquarters' comments attached: 'When we read them, we nearly wept. Not a single individual had had the courage openly to support our suggestion, not even those who had privately given it their cordial approbation.' In the spring of 1915 Sir Eric Geddes and a machine-gun specialist from the Ministry of Munitions visited France in an effort to discover just what were the Army's anticipated requirements over the next months. They visited Baker-Carr's School at Wisques and asked him how many machine guns he thought might be needed. He said that he thought 20,000 might just about be sufficient. Geddes asked GHQ for their reactions to this figure. They treated the whole thing as a rather bad joke: 'I was given to understand that steps would be taken to put the matter right, and a reasonable number of guns would be ordered.'

As well as advocating a substantial increase in the number of machine guns, Baker-Carr was also one of the first to press for the creation of a separate machine gun corps. This idea was at first no better received than the other. His memoranda on the subject were put into a box at GHQ labelled 'Of No Further Interest'. Summing up his experiences in trying to arouse that interest, Baker-Carr has written:

For six months and more I had fought everyone at GHQ to get my scheme through and, except for my own staff, I had never received one word of help or encouragement. I had been told that I was a visionary, a fanatic, a meddler with things that did not concern me, an insubordinate young 'pup' and several other complimentary names. My scheme had been characterised as ridiculous, impossible, impracticable, subversive . . . and contrary to all accepted Military Practice.[11]

Tactical Preconceptions

The most tragic aspect of the High Command's blindness to the power of automatic weapons was their chronic inability to recognise exactly what it meant to face up to an enemy equipped with such weapons. For all power had now passed to the defensive and the main reason for this was the presence of the machine guns. As one writer has put it: 'It was as simple as this: three men and a machine gun can stop a battalion of heroes.'[12] The essential tragedy of the First World War is that the British commanders did not grasp this basic fact over three years.

A French machine-gun crew in 1915

The war on the ground had congealed almost immediately. The first trench warfare took place in September 1914 as the Germans on the Chemin des Dames Ridge, on the Aisne, dug in to block the Allied advance. By October 1914 one officer was able to write that: 'The foremost infantry of both armies are now too securely entrenched, 200 to 400 yards apart, for attacks to have much chance of success save at prohibitive price.'[13] These were the conditions that prevailed for the next three years and more. The power of the defence, and particularly the machine gun, had rendered almost nil the chances of a successful frontal attack. But the High Command remained resolutely oblivious to this fact. For the generals:

> Battles were not going to be won by subtlety and manoeuvre; the decision would go to the commander who displayed the greatest moral fibre, who, undismayed by his own casualties, forced the enemy to exhaust his reserves until the moment came when nerve and resilience snapped, the battle-line broke and the victor was able to surge irresistibly forward to dictate peace in the enemy's capital . . . Few commanders doubted that the pattern of Austerlitz, of Jena, of Sadowa and Sedan would be repeated, on a Brobdignagian scale.[14]

The soldiers' drill and exercises, for example, showed what kind of battle the generals expected. Even at the end of the previous century intelligent soldiers like Meinertzhagen were complaining that close-order drill had very little relevance to the conditions of modern warfare. The first few months of the First World War showed that it was as much a relic of the past as the longbow. But no one seemed capable of absorbing the lesson. After the decimation of the British Expeditionary Force it was realised that a vast new army would have to be raised. By 15 September 1914 some 700,000 men had been enlisted. In administrative terms it was a considerable achievement, but in terms of training and preparedness the men went like lambs to the slaughter. For one such battalion in late 1914:

> The training programme . . . was still mainly concerned with field tactics of the kind used in the Boer War. The whole battalion would form sections and deploy in open order, each company led by its commander and the colonel at the head of the battalion. They were taught to form section, advance and deploy in echelon, moving forward all the time and finally

pushing home a mock attack at the point of the bayonet.

Artist's attempt to show that the man was mightier than the machine

The men of this battalion, of the York and Lancaster Regiment, were also of course given rudimentary musketry training, and over the months they began 'to wonder what might happen to long lines of slowly moving infantry if they were faced with a determined and well-equipped opposition. One

technically minded ex-pit deputy pointed out that the German Mauser rifle was sighted up to 600 yards and the combined length of his rifle and bayonet was something under six feet. He than indulged in some lively speculation as to what might happen when he attempted to cover the odd 598 yards.'[15] It was well for this soldier's relative peace of mind that he was not very aware of the comparative merits of the German machine guns.

But this kind of training went on throughout 1915 and beyond. A member of an East Anglian battalion wrote of his training: 'A few lessons on how to dig and drain trenches, on raiding, fixing up barbed wire, using Very lights, caring for trench feet, using a machine gun, dodging mortar bombs . . . wouldn't have been amiss. Drilling, marching, musketry, physical jerks, advancing in artillery formation, and a few other open warfare manoueuvres about made up our training.'[16] The same officer who had noted in October 1914 that open warfare was impossible also made the following observations about training in the months to follow. In January 1915 he noted in his diary: 'The pre-war daily routine is carried out . . . It includes two or three hours of close-order drill, musketry, bayonet exercise . . . ' Nine months later, 'We are busy . . . practicing our old pre-war methods of attack, i.e. deploying from advancing platoon and section columns into extended firing line and supports to close with the enemy.'[17]

Hand in hand with this obsession with obsolete attacking formations went a fierce emphasis upon the bayonet. For the men of the York and Lancaster Regiment: 'Another major feature of their training was bayonet drill. "The bayonet," their colonel told them, "is the ultimate weapon in battle"; a doctrine which was passed on by their senior NCOs and implicitly believed by the troops.'[18] The army even had a special officer, a Highlander, whose sole duty was to lecture to the troops on the use of the bayonet. Even in the summer of 1916, during the preparations for the Battle of the Somme, Major (later Colonel) Campbell's pep-talks were regarded as an indispensable part of battle training. Siegfried Sassoon said of him at this period:

> He spoke with homicidal eloquence, keeping the game alive with genial and well-judged jokes. Man, it seemed, had been created to jab the life out of the Germans. To hear the Major talk one might have thought that he did it himself every day before breakfast. His final words were: 'Remember that every Boche you fellows kill is a point scored to our side; every Boche you kill brings victory one minute nearer. Kill them! Kill them!'[19]

Ironically, for Sassoon was bitterly opposed to the war, Campbell's melodramatics do seem to have had some effect. After hearing this lecture he wrote a poem called *The Kiss*, which begins thus:

To these I turn, in these I trust –
Brother Lead and Sister Steel.
To his blind power I make appeal,
I guard her beauty clean from rust.

As Sassoon said later: 'I originally wrote it . . . after being disgusted by the barbarities of the famous bayonet-fighting lecture . . . The difficulty is that it doesn't show any sign of satire.'[20] But if Campbell had actually stirred such an intelligent man to unintentional militarism, it merely serves to underline the folly of giving the ordinary soldier the impression that the heavily fortified German lines could be carried with such ease. For, to use the Major's own terminology, there were few 'points' scored in the offensive on the Somme. The German machine guns made sure that all but a handful of men never got near enough to stick their bayonets into anybody. But were either he or the High Command downhearted? Not a bit of it. In the preparations for the Third Battle of Ypres, or Passchendaele as it is more commonly known, the good Major, newly promoted, was at it again. General Jack tells us, in July 1917: 'The day before yesterday a bloodthirsty fellow, Colonel Campbell, the Army bayonet-fighting expert, gave a lurid lecture . . . on the most

Mowing the Hun down at Tilloy April 1917

economical use of the bayonet, to arouse the pugnacity of the men.'[21] In this last purpose he may well have succeeded, but it availed the attackers nothing. Again the machine guns mowed them down.

As a final example of the generals' total sense of unreality during the First World War one must cite their continued faith in the value of cavalry. As has been seen in an earlier chapter, it should have been clear by 1905, at the very latest, that modern firepower had rendered cavalry almost useless. 'In the Russo-Japanese War an English observer, the future General Sir Ian Hamilton, reported that the only thing the cavalry could do in the face of entrenched machine guns was to cook rice for the infantry.' Unfortunately the only effect of such blunt realism was to make 'the War Office . . . wonder if the months in the Orient had not affected his mind.'[22] It was to take more than a war conducted on the other side of the world to shake the average British officer's faith in the ultimate decisiveness of the cavalry charge.

In 1913 one officer was given a depressing preview of the bankruptcy of future tactical thinking. Describing a conversation with the future Commander-in-chief, he relates that: 'I asked Haig . . . why there were four brigades in the cavalry division, more than any one man could control, as the Germans had discovered. He replied: "But you must have four." "Why?" *"For the charge."* "Two brigades in the first line, one in support, and you must have one in reserve." '[23] The years that followed witnessed many desperate efforts to translate this vision into reality. During the retreat from Mons in the first of the war, the 9th Lancers and the 18th Hussars were mown down by the German machine guns as

Rearguard Italian action on the Isonzo 1917

they attempted a flank attack near Valenciennes. But the army as a whole was not impressed. Writing in his diary in October 1914, General Jack noted that: 'Our advanced cavalry always ride sword in hand or lance at the "carry", and charge at sight any hostile mounted bodies within charging distance.'[24] In June 1916, after nearly two years in which the cavalry had consumed vast quantities of forage and achieved virtually nothing, Haig had an interview with the King at Buckingham Palace. At one stage the King politely drew Haig's attention to the financial burden imposed by all these useless horsemen. As Haig described it: 'He thought that cavalry should be reduced on account of the cost of maintenance. We could carry on the war for a very long time provided the cost did not exceed £5,000,000 a day. I protested that it would be unwise, because in order to shorten the war, and reap the fruits of any success, we must make use of the mobility of the Cavalry.' In the early evening of the 14th July, Haig put his theories to the test. Two squadrons each of the 20th Deccan Horse and the 7th Dragoon Guards carried out a charge against unshaken German infantry in High Wood. As they came out of the cornfields in front of the wood a German machine gun opened up and the attackers were forced to retire. As many eye-witnesses testified afterwards, it was certainly magnificent. One would like to complete the cliché and say that even so it was not war. But the tragedy is that for those who had ordered this futile gesture it was the very essence of the correct offensive spirit. As the German commander in this sector wrote: 'The frontal attacks over open ground against a portion of our unshaken infantry, carried out by several English cavalry regiments, which had to retire with heavy losses, gives some indication of the tactical knowledge of the Higher Command.'

This knowledge was not in the least enriched by the débâche at High Wood. In September 1916 Lloyd George himself went to France to confer with his generals about preparations for the year's last attacks. During this visit he noted:

> I have driven through **squadrons** of cavalry clattering proudly to the front. When I asked what they were for, Sir Douglas Haig explained that they were brought up as near to the front line as possible, so as to be ready to charge through the gap which was to be made by the Guards in the coming attack. The cavalry were to exploit the anticipated success and finish the German rout . . . When I ventured to express my doubts as to whether cavalry could ever operate successfully on a front bristling for miles behind the enemy's lines with

barbed wire and machine guns . . . the Generals fell on me.[25]

Of course the chance never came. The first infantry attacks ground to a halt long before there was ever any possibility of sending in the cavalry to exploit a break in the German line.

To the generals this was merely a temporary setback. It most certainly did not indicate any fundamental change in the character of modern warfare. In April 1917 Allenby, later the hero of the Palestine campaign, but at this time still serving in France, wrote regretfully to his wife: 'The cavalry nearly got a chance yesterday, but it did not quite come off. Wire and machine guns stopped them . . . They suffered somewhat severely.'[26] On 20 July 1917 General Jack was still able to write in his diary: 'High Command . . . continue to expect that infantry assaults will burst a gap in the German defences large enough for horsemen to ride through . . . The 10th Hussars, which lost some two-thirds of their men at the Battle of Arras last spring, do not appear to share this belief.'[27] Yet in June 1918 a British machine gunner was still able to write in his diary: 'Not far from us . . . were the remains of Monchy Le Preux, where the cavalry had been cut up in the early stages of the battle. My pals told me that the cavalry attack was a fiasco. It is impossible to understand the reason for throwing . . . cavalry against machine guns skilfully emplaced behind a screen of barbed wire.'[28]

But there is much in the First World War that defies understanding. Nor were the British generals the only ones whose actions seem inexplicable. In May 1916 a French infantry officer was also treated to the sight of a regiment of lancers forming for the attack. One of his fellow officers turned to him and said resignedly: 'They're holding back all these fellows for the breakthrough, the famous breakthrough that we've been waiting for for two years . . . You know there's nothing like a lance against machine guns.' But at least these particular cavalry men had horses. In June 1918 the same French officer was part of a unit that was called upon to relieve another at the front. The men departing turned out to be a troop of dragoons, who told how they had been ordered to attack the German lines the day before. 'They added that, in accordance with the orders received to maintain the cavalry spirit, they had charged on foot with their lances at the ready.'[29]

The Machine Gun Triumphant

These then were the symptoms of the hidebound attitudes of

130

the Allied High Commands, attitudes centred around the old notion of the glorious charge and the breakthrough, and relying upon the timely use of the bayonet, the sabre and the lance. It remains now to see what were the consequences of this inanity.

One of the first British offensives took place at Neuve Chapelle, in March 1915. The offensive was doomed to failure from the first because it was carried out with insufficient men and equipment. It had been intended that the French and the British should share the burden but Sir John French had tired of ceaseless French demands for aid in other sectors, and in a fit of pique decided to mount the attack with his own forces. The events of the first day made abundantly clear the characteristics of the new style of warfare. The actual infantry assault was preceded by a concentrated artillery barrage, intended to knock out the front line defences. When it stopped the British were to move forward. The experiences of one unit, the 2nd Scottish Rifles, in this battle have been meticulously studied, and are typical of the day's events:

> Ferrers was first out from 'B' Company, his monocle in his eye and his sword in his hand. As the guns stopped firing there was a moment of silence. Then the guns started again, firing behind the German lines . . . Almost at the same moment came another noise, the whip and crack of the enemy machine guns opening up with deadly effect. From the intensity of their fire, and its accuracy, it was clear that the shelling had not been as effective as expected . . . As the attack progressed the German positions which did most damage were two machine gun posts in front of the Middlesex. Not only did they virtually wipe out the 2nd Middlesex with frontal fire, but they caused many of the losses in the 2nd Scottish Rifles with deadly enfilade, or flanking fire.[30]

In other words, two machine guns, a dozen Germans at

German Maxims as anti-aircraft guns

most, brought to a halt two battalions of British infantry, or something over 1,500 men. Similar experiences were recorded all down the line of attack, and as the Germans pushed more men and machine guns into the threatened sector the attack swiftly fizzled out.

But the Allied High Command drew no lessons from this reverse. It was merely felt that the next time all that was needed was more shells, more men, and possibly just a little more offensive spirit. In September, at the Battle of Loos, exactly the same thing happened again. But still no one thought to question the British tactics. Instead they got rid of the commander-in-chief. At the end of the year French was replaced by Haig.

The basic result of this change was that nothing changed. Haig went ahead with plans for a massive offensive of his own. Cleverly he chose to attack on the Somme, possibly the strongest section of the German line. Obviously the extensive preparations needed for a full-scale offensive could not be hidden from the Germans, so they were able to make their defences even stronger. The key component of these defences were the machine guns. The Germans put much careful thought into the problem of how to utilise them most effectively. In June orders were issued to all machine gunners:

A Russian machine gun post in 1915

When siting machine guns in the front line, it must be remembered that it is possible for the enemy in an attack on a large scale to force an entrance into the front line before the machine guns can come into action. This can only be avoided by siting the emplacement not on the parapet but *behind* the parados of the front line: this method at the same time affords a much better field of fire. In addition, owing to the feeling of safety which this position inspires, the men will work their guns with more coolness and judgement.[31]

During May and June the machine gunners constantly practised bringing up their weapons from their deep dug-outs and getting them into action in the minimum of time. Eventually they were able to emerge from the complete safety of the dug-outs and have the guns assembled and firing in under three minutes. They were taught to train their guns on those parts of the protective wire that might have been cut by the preliminary artillery bombardment, and other crews were kept in reserve to cover the more serious gaps.

One cannot be sure how much of this preparation was known to the British commanders. Yet the experiences of 1915 should surely have indicated that a crude frontal assault could have little real chance of success. In that year an officer at the front had written to the Northcliffe press to point out that: 'Those who have had a little experience . . . know that over some 200 yards it is only machine guns that can stop them. When the first line has been destroyed by them . . . many of us have seen the second immediately throw themselves over our parapet to certain and conscious destruction.'[32] Yet in May, the *Tactical Notes* issued by the Fourth Army headquarters prescribed that when the infantry went forward on the day of the attack: 'The leading lines should not be more than 100 yards apart, with the men in each line extended at two or three paces interval.' And GHQ's instructions stipulated that: 'The assaulting columns must go right through above ground to the objective in successive waves or lines.' As a final absurdity every man was required to carry with him equipment which altogether weighed a little under 70lbs. and reduced his maximum speed to almost a walking pace. As the *Official History* noted:

> The total weight . . . made it difficult to get out of a trench, and impossible to move quicker than a slow walk, or to rise and lie down quickly. This overloading of the men is by many infantry officers regarded as one of the principal reasons for the heavy losses and failure

of their battalions; for their men could not get through the machine gun zone with sufficient speed.[33]

On 1 July 1916 the British battalions were sent over the top. The pattern of events was clear from the very beginning. The artillery bombardment had obviously had no effect on the deep German dug-outs.

> Although the attack began at 7.30 a.m., even before that time not only was the battalion [York and Lancaster] beginning to suffer casualties from shell-fire, but machine guns firing from the German second and third lines were traversing along the top of the trench. Gilbert remembered looking up and seeing the coarse green grass growing along the parapet; flying into the air as the bullets' stream hit it. It was just like the effect produced by a lawn mower being used without a grass box.[34]

But then the word was given that the men had to leave the comparative safety of the trenches and face the machine guns in the open. The following description of what followed, by a German machine gunner, shows how ill-founded were the German fears of a breakthrough into the first line of trenches:

> When the English started advancing we were very worried; they looked as though they must overrun our trenches. We were very surprised to see them walking, we had never seen that before . . . The officers were in front. I noticed one of them walked calmly, carrying a walking stick. When we started firing we just had to load and reload. They went down in their hundreds. You didn't have to aim, we just fired into them.[35]

Jack's battalion was there too, and like every other one it suffered terribly from the machine gun fire: 'The enemy's machine guns some 1,400 yards from my position, now swept the crest like a hurricane and with such accuracy that many of the poor fellows were shot at once. This battalion had 280 casualties in traversing the 600 yards from our front line.'[36]

Other battalions suffered even more heavily. Two of them, the 7th Green Howards and the 10th West Yorkshires, had been designated to attack the heavily fortified village of Fricourt. Both battalions were practically wiped out within three minutes by a single cunningly situated machine gun. A similar fate awaited the 16th Northumberland Fusiliers who

Germans and Maxim
in Poland 1915

135

were to storm the ruins of the village of Thiepval. Two companies were sent forward behind a rugby ball, drop-kicked from the assembly trench. They rushed for the gaps in the wire but found that the German machine gunners trained their guns on them immediately. In a matter of minutes the two companies, faced by four machine guns, had been reduced to eleven men. A.H.Farrar-Hockley has described, in a chilling matter-of-fact way, the assault of 88 Brigade, part of the 29th Division, on Y Ravine:

> The long burst of a machine gun does not kill a battalion; indeed, some men in a line will almost certainly pass through the run of bullets, in the gaps between them, so to speak. But if the line of men perseveres with a determined gallantry, over a long open approach, the end is certain. So it was before Y Ravine. The Newfoundland officers and men would not halt; they had orders to advance into the enemy line: they advanced. 710 men fell. Some minutes later, three companies of the Essex emerged on their own front to fall, as gallantly but as forlornly just inside the enemy line.'[37]

A similar fate befell almost every one of the 143 battalions that attacked that day. Fifty per cent of these men became casualties, as did 75 per cent of their officers. The German line was not broken, hardly any ground was taken, and even less was held for more than a few hours.

It was principally the machine guns that had caused this carnage. In his dispatch from the Somme on 2 July, Philip Gibbs wrote:

Indians carrying Vickers guns

It was the fire of the German machine guns which was most trying to our men. Again and again soldiers have told me today that the hard time came when these bullets began to play upon them. In spite of our enormous bombardment there remained here and there, even in a front-line trench, a machine gun emplacement so strongly built . . . that it had defied our high explosives . . . The same opinion . . . was given me today by many men whose bodies bore witness to these German Maxims . . . 'It seemed to me', said a Lincolnshire lad, 'as if there was a machine gun to every five men.'[38]

The German Official History is also full of praise for the work of the machine gunners. It tells of 'the wonderful effect of the machine guns' and relates how the British advanced in 'solid lines without gaps in faultless order, led by . . . officers carrying battle flags and sticks. Wave after wave were shot down by well-aimed fire . . . A wall of dead British was piled up on the front.'[39] One of the most moving descriptions of all was given by Edmund Blunden, many years later, in his *War Poets* collection:

> And everywhere I see the faces and figures of enslaved men, the marching columns . . . the files of carrying parties . . . the 'waves' of assaulting troops lying silent and pale on the tape-lines of the jumping-off places.
> . . . I go forward with them . . . up and down across ground like a huge ruined honeycomb, and my wave melts away, and the second wave comes up, and also melts away, and then the third wave merges into the ruins

A motorised section of the Machine Gun Corps 1916

137

of the first and second, and after a while the fourth blunders into the remnants of the others, and we begin to run forward to catch up with the barrage, gasping and sweating, in bunches, anyhow, every bit of the months of drill and rehearsal forgotten.

We come to wire that is uncut, and beyond we see grey coalscuttle helmets bobbing about . . . and the loud crackling of machine guns changes to a screeching as of steam being blown off by a hundred engines, and soon no one is left standing . . . The brigade, with all its hopes and beliefs, has found its grave on those northern slopes of the Somme battlefield.[40]

Amazingly enough the very soldiers who had suffered so cruelly at the hands of the German machine gunners still felt impelled to exhibit their British sense of 'fair play' and give these gunners full credit for their remarkable heroism. For the intensive German training had most certainly succeeded in making them a redoubtable *corps d'élite*. Lieutenant-General Carton de Wiart, who had fought on the Western Front, had this to say about the machine gunners on the Somme:

The German machine gunners were outstanding, almost invariably very brave men and the pick of the German Army. One day on the Somme we were held up for a considerable time by a German machine gun. Finally we silenced it and were able to advance, and found the whole crew dead, but all of them bandaged, having been wounded several times before being killed. Another time we found a young German lying dead beside a machine gun, and the villagers told us that all the Germans had retired except this young boy who had remained firing his gun until he was killed.[41]

There were also spontaneous tributes at the time. Whilst Philip Gibbs was interviewing the survivors of the first day of the battle:

A young officer of the Northumberland Fusiliers paid. a high tribute to them. 'They are wonderful men . . . and work their machines until they are bombed to death. In the trenches by Fricourt they stayed on when all the other men had either been killed or wounded, and would neither surrender nor escape. It was the same at Loos, and it would not be sporting of us if we did not say so, though they have knocked out so many of our best.'[42]

Sir Henry Newbolt, and no doubt every public school master in England, would have been proud of a man who could think in terms of what was and was not 'sporting' in the midst of so much horror. It is certain that those back in England still thought that the war was one in which such values were still relevant. Captain A.J.Dawson wrote a book called *Somme Battle Stories* in which one officer reputedly told him: 'This business of fighting – fighting continuously and cheerily in the presence of devastating casualties – has a good deal in common with swimming and bicycling and things of that sort in which instinct plays a big part; and horse-riding too.'[43]

The troops at the front knew better, though not unfortunately the generals. Even after the bloody futility of the first few days, Haig kept the offensive going for a further three months, throwing men against German positions that were continually being reinforced. The Anzac Corps figured prominently in these later assaults. On the strategic rationale behind this chronic blood-letting, the Australian Official History said:

> . . . to the front line the method merely appeared to be that of applying a battering-ram ten or fifteen times against the same part of the enemy's battle front with the intention of penetrating for a mile, or possibly two, in the midst of his organised defences . . . Even if the need for maintaining pressure be granted, the student will have difficulty in reconciling his intelligence to the actual tactics.

The History goes on to quote the remarks of some of the men who were present. Said one: 'For Christ's sake, write a book on the life of an infantryman, and by so doing you will quickly prevent these shocking tragedies.' In his last letter before he was killed a lieutenant wrote of the 'murder' of his friends 'through the incompetence, callousness, and personal vanity of those high in authority.' After one spell in the trenches another officer wrote: 'We have just come out of a place so terrible that . . . a raving lunatic could never imagine the horror of the last thirteen days.'[44]

British casualties on the Somme were over a quarter of a million men, and nowhere along the twelve-mile front was more than eight miles of territory gained. By any conventional standards the whole campaign had been an appalling failure. Yet in 1917 Haig resolved to attempt exactly the same sort of offensive. His final dispatch from the Somme, on 23 December, had included the observation that: 'Machine guns play a great part – almost a decisive part under some conditions – in modern war.' Yet in the following year he quickly fell in with Nivelle's plan for another major offensive when these very conditions applied as much as ever. The basic strategical assumptions behind this offensive, in Champagne, were almost mystical in character. According to Nivelle: 'The character of violence, of brutality and of rapidity must be maintained. It is in the speed and surprise caused by the rapid and sudden eruption of our infantry upon the third and fourth positions that the success of the rupture will be found. No consideration should intervene of a nature to weaken the *élan* of the attack.' Chief among such considerations was common sense. The British part in this offensive was limited to attacks around Arras and Bullecourt. The first was relatively successful although it soon bogged down. The second, by Gough's Fifth Army, came to

Captured German machine-gunners 1917

be '. . . employed by British instructors afterwards as an example of how an attack should not be undertaken.' The French attack itself was torn to pieces by the German machine guns, and the *poilus* came near to general mutiny. Their attitude was typified by remarks such as: 'We are not so stupid as to march against undamaged machine guns.'

For Haig all this was simply a sign that the British must now undertake a major offensive of their own. As Sir Henry Wilson described it, it was to be 'a Somme with intelligence'. It was certainly another Somme, but it would take an assiduous researcher to find much evidence of any intelligence. As before, British tactics were based upon the infantry charge, and the German upon the deployment of large numbers of machine guns. The forward positions were almost entirely entrusted to the battle-hardened machine gunners, whilst the bulk of the ordinary infantry was held further back to counter-attack or mop up the few enemy who managed to get beyond the front line. The results too were exactly the same. As Lloyd George put it:

Members of a machine-gun company in 1917

> Our men advanced against the most terrible machine gun fire ever directed against troops in any series of battles, and they fell by the thousands in every attack. But divisions were sent on time after time to face the same slaughter in their ranks, and they always did their intrepid best to obey the fatuous orders. When

divisions were exhausted or decimated, there were plenty of others to take their places.[45]

Through bitter experience the machine taught that man himself was no longer master of the battlefield. The individual counted for nothing, all that mattered now was the machinery of war. If a machine gun could wipe out a whole battalion of men in three minutes, where was the relevance of the old concepts of heroism, glory and fair play between gentlemen? Lloyd George said that almost eighty per cent of the First World War casualties were caused by machine guns. In a war in which death was dealt out to so many with such mechanical casualness how could the old traditional modes of thought survive? The First World War was an event of crucial significance in the history of Western culture, a four-year trauma in which men tried to hold on to their old self-confidence in the face of horrors that would have been literally unimaginable two or three years earlier.

The confusion manifested itself in many ways. Some tried to cling on to the old modes of thought. In this respect the praise for the German machine gunners achieves a new significance. In it one sees a desperate attempt to create new heroes in a war in which heroism was in fact irrelevant. Men tried to forget the weapons themselves, the mere machines that killed so unerringly and so indiscriminately, and remember only the men that pressed the trigger. Thus death could be made a little more acceptable.

But there was also a new brutalisation. One can detect this particularly in the machine gunners themselves. In his history of the Machine Gun Corps, Hutchison recounts his own experiences as a member of that body. Recalling an attack on High Wood, during the Battle of the Somme, he tells how his company was massacred almost to a man:

> With my runner I crept forward among the dead and wounded, and came to one of my guns mounted for action, its team lying dead beside it. I seized the rear leg of the tripod and dragged the gun some yards to where a little cover enabled me to load the belt through the feed-block. To the south of the wood Germans could be seen, silhouetted against the sky-line, moving forward. I fired at them and watched them fall, chuckling with joy at the technical efficiency of the machine.

A little later German artillery began firing and the shells fell amongst the British wounded. 'Anger, and the intensity of the fire, consumed my spirit, and not caring for the consequences, I rose and turned my machine gun upon the battery,

laughing loudly as I saw the loaders fall.' In another incident, in April 1918, Hutchison was a member of a machine-
gun battalion that played a big part in holding back a part of
the German offensive of that year. His description of it gives
a telling example of how men had been numbed by the years
of slaughter and had made for themselves new standards of
beauty:

> The particular incident . . . is probably the most
> thrilling in which organised machine gunners have ever
> participated. The rapidity of action; the extraordinary
> situation; the perfect discipline and drill; the setting of
> untouched farmhouses, copses and quietly grazing cat-
> tle; the flying civilians with their crazy carts piled high
> with household chattels and the retreating infantry
> behind; the magnificent targets obtained . . .

*A Lewis Gun
adapted for aerial
warfare*

One is reminded more of the Viking beserkers than of the officers and gentlemen who had gone to France in the early days. Nor was their fanaticism turned against the enemy alone. On that same day Hutchison and his adjutant 'discovered in the Belle Croix *estaminet* behind the mill a crowd of stragglers, fighting drunk. We routed them out, and, with a machine gun trained on them, sent them towards the enemy. They perished to a man.'[46]

Even those who retained their humanity and sensitivity also show signs of the spiritual confusion that overtook those on the Western Front. Nowhere is this more clearly evidenced than in the work of the First World War poets. A poem of Siegfried Sassoon's, *The Redeemer*, shows this as well as any one poem can. Sassoon tells how he saw an English soldier struggling along with a load of planks, and in the darkness imagined for a moment that he was Christ on his cross. The last verse reads:

> He faced me, reeling in his weariness,
> Shouldering his load of planks, so hard to bear.
> I say that he was Christ, who wrought to bless
> All groping things with freedom bright as air,
> And with his mercy washed and made them fair.
> Then the flame sank, and all grew black as pitch,
> While we began to struggle along the ditch;
> And someone flung his burden in the muck,
> Mumbling: 'O Christ Almighty, now I'm stuck!'

The intentionally pathetic ending clearly reveals Sassoon's feeling that, try as he might to rekindle some belief in the old religious and ethical sureties of the days before the war, a new reality has been born in which such beliefs count for nothing besides the facts of the mud, the darkness and the heavy load. One critic has put the whole thing very well:

> The poets of World War I made it clear that man could no longer depend on his personal courage or strength for victory or even survival; mechanisation, the increased size of armies, the intensification of operations, and the scientific efficiency of long-distance weapons destroyed the very elements of human individuality: courage, hope, enterprise, and a sense of the heroic possibilities in moral and physical conflict.

Yet this critic goes on to upbraid the poets of the First World War for being too wrapped up in their individual suffering. He accuses them of 'a lack of historical perspective which in World War I poetry resulted in a lack of temporal depth, since any event ceases to have a real significance unless it is in some way related to other events.'[47] But surely the whole point of this poetry is that there was no chance of a historical perspective? How could it be related to other events when there had never been anything remotely like it? Those poets wrote to express their horror of a war that they could hardly comprehend as a meaningful part of the historical process. The horror and the confusion are the enduring message in what they wrote. For them the war had no meaning and the ideals that had sustained them in the beginning had become an irrelevancy. The poets felt that they were living on the brink of Chaos. One can hardly blame them for having described it so well.

This was the effect of the war on almost everyone that took part in it, even as an observer. The military writer, Repington said in 1914: 'It transcends all limits of thought, imagination and reason. We little creeping creatures cannot see more than a fraction of it . . . We look, gasp, wonder and are dumb. We do not know. Nobody knows. This war, for once, is bigger than anybody. No one dominates it. No one even understands it. Nobody can.'[48] It was D.H.Lawrence who wrote the epitaph of the old world when he spoke of 'the terrible, terrible war made so fearful because in every country practically every man lost his head, and lost his own centrality, his own manly isolation in his own integrity, which alone keeps life real.'[49] Perhaps this new conception of men as mere units before the might of the machine gun was never better expressed than upon the Derwent Wood memorial to the Machine Gun Corps. It is a statue of 'The Boy David' and still stands at Hyde Park Corner. Its inscription reads:

Saul hath slain his thousands
But David his tens of thousands.

Notes

1. de la Gorce, op.cit., p.55.
2. J.Terraine (ed.), *General Jack's Diary*, Eyre and Spottiswoode, London, 1965, p.55.
3. Sir L. Woodward, *Great Britain and the War of 1914-18*, Methuen, London, 1968, p.36 (footnote).
4. Luvaas, *Education*, op.cit., p.313.
5. Wahl and Toppel, op.cit., p.14.
6. A.Farrar-Hockley, *The Somme*, Pan Books, London, 1966, p.193.
7. Baker-Carr, op.cit., p.72.

8. Clarke, op.cit., p.106.
9. Ibid, p.65.
10. D.Lloyd George, *War Memoirs*, Odhams Press, London, 1938, pp.359–60.
11. Baker-Carr, op.cit., pp.82, 85, 87, 122 and 135.
12. G.Blond, *Verdun*, Mayflower-Dell, London, 1967, p.23.
13. Terraine, op.cit., p.55.
14. M.Howard, *Studies in War and Peace*, Temple Smith, London, 1970, p.107.
15. R.Haigh and P.Turner, *Not For Glory*, Robert Maxwell, London, 1969, pp.18 and 25.
16. B.Gardner, *The Big Push*, Sphere Books, London, 1968, p.52.
17. Terraine, op.cit., pp.90 and 111.
18. Haigh and Turner, op.cit., p.17.
19. Gardner, op.cit., p.52.
20. I.Parsons (ed.), *Men Who March Away*, Heinemann, London, 1965, p.17.
21. Terraine, op.cit., p.227.
22. B.Tuchmann, *August 1914*, Four Square Books, London, 1964, p.216.
23. Gardner, *Allenby*, op.cit., pp.75-6.
24. Terraine, op.cit., p.63.
25. Gardner, *Push*, op.cit., pp.40, 115 and 133-4.
26. Gardner, *Allenby*, op.cit., p.107.
27. Terraine, op.cit., p.229.
28. G.Coppard, *With a Machine Gun to Cambrai*, HMSO, London, 1969, p.110.
29. R.Arnaud, *Tragédie Bouffe*, Sidgwick and Jackson, London, 1966, pp.53 and 145.
30. J.Baynes, *Morale: the 2nd Scottish Rifles at the Battle of Neuve Chapelle, 1915*, Cassell, London, 1967, pp.68 and 71-2
31. Hutchison, op.cit., p.149.
32. Lloyd George, op.cit., p.127.
33. Gardner, *Push*, op.cit., p.72.
34. Haigh and Turner, op.cit., p.41.
35. M.Middlebrook, *The First Day on the Somme*, Allen Lane, London, 1971, p.157.
36. Terraine, op.cit., p.146.
37. Farrar-Hockley, op.cit., pp.122-3.
38. Sir P.Gibbs, *The War Despatches*, Times Press, London, 1964, pp.109-10.
39. Gardner, *Push*, op.cit., p.104.
40. E.Blunden, *War Poets 1914–18*, The British Council and the National Book League, London, 1958, p.13.
41. Lt.-Gen. Sir A.Carton de Wiart, *Happy Odyssey*, Pan Books, London, 1955, p.55.
42. Gibbs, op.cit., p.109.
43. Gardner, *Push*, op.cit., p.158.
44. Liddell-Hart, *First World War*, op.cit., p.249.
45. L.Wolff, *In Flanders Fields*, Pan Books, London, 1961, pp.78. 82, 85, 95, 205.
46. Hutchison, op.cit., pp.168 and 246.

47. J.H.Johnston, *English Poetry of the First World War*, Princeton University Press, Princeton, 1964, pp.10, 14 and 16.
48. Luvaas, *Education*, op.cit., p.328.
49. G.Panichas (ed.), *Promise of Greatness*, Cassell, London, 1968, pp.xxii-xxiii and xxix.

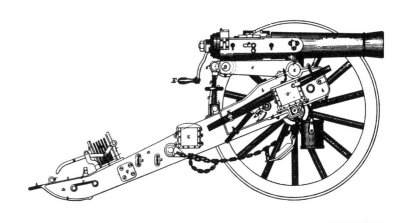

VI *A Symbol of the Times*

NICK: You . . . you swing wild, don't you?
MARTHA: Hah!
NICK: Just . . . anywhere.
MARTHA: Hah! I'm a Gatling gun.
Hahahahahahahahahaha!
NICK: Aimless . . . butchery. Pointless.

Edward Albee: *Who's Afraid of Virginia Woolf?*

After 1918 the machine gun ceased to make headlines as a military weapon. However, though it had lost its battlefield dominance, one particular version of it was destined to hit the headlines again on the streets of a supposedly peaceful world. For in the summer of 1918 the first working model of the Thompson sub-machine gun had appeared.

The gun had been designed by Colonel J.T.Thompson specifically as a weapon for trench warfare, and he dubbed it the 'trench broom', to be used in close-quarter combat when troops were trying to clear an enemy trench. In August 1916 he set up a special company, the Auto-Ordnance Corporation of New York, to develop the gun. But the war ended before it could be put into production, though half a million dollars had already been spent on it. Nevertheless Thompson tried to interest the military in his gun. Both the Army and the Marine Corps carried out tests in 1920 and 1921, but very few orders were made.

Indeed, throughout the 1920s and 1930s the record of the Auto-Ordnanace Corporation was a very poor one. Basically no one had much need for sub-machine guns. After his failure with the military Thompson tried all kinds of ploys to interest other groups in his weapon. First he tried the police. In 1922 the company offered a special cartridge containing bird-shot, which, it was claimed, 'would be useful to authorities in dealing out a lesser degree of punishment. They allow serious occasions and disorders to be handled by officers of the law in the most humane manner possible.' In May of that year the gun was demonstrated to police and journalists at Tenafly, New Jersey. It was predicted by the company, with some justification, that the weapon would 'reform or remove bandits instantly.'[1] A few sales were made to police forces in New York, Boston, San Francisco, and the states of Pennsylvania, Massachusetts, West Virginia, Connecticut and Michigan. But in most cases the guns were used not so much to deal with criminals as to intimidate striking workers. In 1920 one writer gleefully outlined their advantages in this respect:

> There would be no trouble whatever for one man firing the gun to sweep a street clear from curb to curb, but after all its greatest strength lies in its moral effect. Killing many of the common American sort of mob is unfortunate unless the right ones can be selected for slaughter. Mobs as a rule are composed of ten per cent vicious (the leaders) and ninety per cent fools. Wherefore the dispersing of the crowds without bloodshed is usually desirable. It would be a most vicious and determined aggregation that would stand up to the fire of one of the guns, after a couple of bursts had been fired over their heads.[2]

Thompson next turned his attention to the air force, and designed a two-man plane to which thirty Thompson guns were attached, all of them pointing downwards. The plane was supposed to move backwards and forwards across the area to be attacked rather in the manner of a crop-spraying plane. On its one and only trip the aircraft proved to be impossibly heavy, and when the guns were fired they all jammed on their own deluge of spent cartridges.

As a final resort Thompson even tried to interest private citizens in his weapon. The company's promotion literature envisaged ranchers using it, in the manner of an updated cowboy, and even private householders. As the historian of the Thompson puts it:

> A company that could fancy a cowboy mowing down bandits, or envision a householder pouring machine gun fire into his darkened dining-room in defence of the

family silver, might well have misjudged its markets. For the sub-machine gun was legally available to anyone, and lack of police and military interest made it by default a civilian weapon. And so it came to pass that the Thompson – manufactured in peacetime, sold on the commercial market – was, in a sense, a machine gun for the home.[3]

Commercially, then, the gun was a flop. And so it remained until the outbreak of the Second World War, when the British, and later the Americans, ordered these guns in their tens of thousands. But even had the Second World War never been fought it seems certain that the 'tommy gun'

The Thompson Submachine Gun
The Most Effective Portable Fire Arm In Existence

THE ideal weapon for the protection of large estates, ranches, plantations, etc. A combination machine gun and semi-automatic shoulder rifle in the form of a pistol. A compact, tremendously powerful, yet simply operated machine gun weighing only *seven* pounds and having only *thirty* parts. Full automatic, fired from the hip, 1,500 shots per minute. Semi-automatic, fitted with a stock and fired from the shoulder, 50 shots per minute. Magazines hold 50 and 100 cartridges.

THE Thompson Submachine Gun incorporates the simplicity and infallibility of a hand loaded weapon with the effectiveness of a machine gun. It is simple, safe, sturdy, and sure in action. In addition to its increasingly wide use for protection purposes by banks, industrial plants, railroads, mines, ranches, plantations, etc., it has been adopted by leading Police and Constabulary Forces, throughout the world and is unsurpassed for military purposes.

Information and prices promptly supplied on request

AUTO-ORDNANCE CORPORATION
302 Broadway *Cable address: Autordco* New York City

A rather fanciful advert for the Thompson Gun 1922

would still be just as familiar to us today. For, during the twenties and thirties, one group in America did take to the gun, ironically those very bandits whom the inventor had hoped to reform or remove. The gangsters of the Prohibition era went on to make the Thompson gun famous throughout the world, though Thompson himself never ceased to be appalled at the fact that his gun had fallen into such hands.

It was first used in organised gang warfare on 25 September 1925, when the Frank McErlane and 'Polack' Joe Saltis gang attacked members of the O'Donnell family in Chicago. From that day it became one of the gangsters' standard weapons, and though pistols, sawn-off shot-guns and bombs were also used quite frequently, for the public at large the 'tommy gun' was the *sine qua non* of the authentic hoodlum. *Collier*'s crime reporter was one of the first to be impressed, to say the least, by the gun's potential in the hands of the gangsters. It was, he said:

> The greatest aid to bigger and better business the criminal has discovered in this generation . . . a diabolical machine of death . . . the highest-powered instrument of destruction that has yet been placed at the convenience of the criminal element . . . an infernal machine . . . the diabolical acme of human ingenuity in man's effort to devise a mechanical contrivance with which to murder his neighbour.[4]

It seems likely that it was McErlane's use of the Thompson that persuaded Al Capone that he too would have to get hold of some. After three consecutive attacks by McErlane on the headquarters, cars and persons of assorted rival beer runners Capone began to realise that such a weapon had innumerable advantages. An American writer, in 1931, during the peak of the era of gang violence, outlined some of the advantages of such a weapon:

> The modern gunman being dependent upon his weapons as well as the line of approach has advanced in mechanical ingenuity considerably during the last ten years. It was Hymie Weiss who invented murder by motor, but the popularity of automobiles very soon gave prior place to the machine gun. In the early days of Prohibition it was no uncommon thing for a gangster to be wounded and picked up by the police or conveyed away by his friends. That kind of bungling has grown unpopular. By the use of a machine gun fatalities became more assured, even if they included a handful of the deceased's friends or a waiter.[5]

So, on 10 February 1926, Capone ordered three Thompson guns of his own. On 20 September he was given further reason to admire the potential of such weapons. On that day a motorcade laid on by Weiss, Capone's chief rival at the time, slowly drove past his headquarters at the Hawthorne Inn, in Cicero, and as each car drove past its occupants methodically raked the building with machine-gun fire. The first man to fire had actually used blanks, hoping presumably to draw Capone outside. Luckily for the gang leader one of his men immediately pushed him to the floor, and during the entire shoot-out, in which one thousand rounds were fired, no one was hurt. But the demonstration was not without its psychological impact. Speaking to a reporter friend afterwards, Capone said: 'That's the gun! Its got it over a sawed-off shotgun like the shotgun has it over an automatic. Put on a bigger drum and it will shoot well over a thousand. The trouble is they're hard to get.'[6] Some months later Capone felt it diplomatic to give rather different reasons for the adoption of the sub-machine gun. By this time he had become something of a hero to the American public, and he began to think that his seedy past needed a little conventional respectability. One of his favourite lines was to tell of his exploits with the American Expeditionary Force in France, during the First World War. In fact he had never got past the medical board, but this did not prevent him from dredging up fond wartime memories. Kenneth Allsop describes how he 'on occasions paid tribute to the American Army for planting the seed in his mind of the machine gun as a piece of business equipment. "The sergeant told me that one man with a machine gun was a master of fifty men with rifles and revolvers," he once told a reporter. "You know something? That guy was right." '[7]

But whatever the reasons for Capone's adoption of the machine gun, there is no doubt that he had learned the lesson well. Knowing that the Cicero motorcade had been laid on by Weiss, on 11 October, two Capone machine gunners chopped him down as he was crossing the street to go into a flower shop. From then on the 'tommy gun' was a regular feature of Capone's operations. He built up a small squad of professional assassins, led by 'Machine Gun' Jack McGurn, who were obliged to train rigorously in private gymnasiums and to practise periodically on machine-gun ranges situated in thinly populated parts of Illinois. At least one of these killers seems to have become a little too attached to his machine gun. This was James 'Fur' Sammons who 'was one of the most dangerous killers in bootlegging and labour racketeering. To him human life was of no value. When he was given a machine gun and sent out on a job of murder, guards would be assigned to accompany *him*. It was their task to prevent him from taking pot-shots at pedestrians for amusement.'[8]

Another rival of Capone's was George 'Bugs' Moran, and it was this feud that provoked what is probably the most famous machine-gun incident of all time. On 14 February 1929, in the so-called St. Valentine's Day Massacre, seven of Moran's men were shot down with machine guns by two Capone killers. One of them was 'Machine Gun' McGurn, the other was Fred Burke, a killer specially imported for the job. Two of the seven killed had in fact tried to eliminate McGurn with a machine-gun attack of their own a little earlier. The revenge was ruthless. Disguised as policemen, the executioners lined the seven against a wall and methodically ran lines of bullets across their heads chests and stomachs. One of them, Frank Gusenberg, with fourteen

The scene of the St. Valentines Day Massacre thirty years later

bullets in him, was actually still alive when police arrived on the scene. 'Frank', said a detective, 'in God's name what happened? Who shot you?' Gusenberg's last words were 'Nobody shot me.' Machine guns dominated the case from beginning to end. The man suspected of actually supplying the guns used, one Frank H.Thompson, was at that time wanted for trying his wares out on his wife and her lover back home in Kirkland, Illinois. In February 1936 McGurn himself was machine-gunned to death in a bowling alley. A comic valentine was left beside the body.

The police never took to the Thompson gun with the same enthusiasm as the gangsters. In 1929, for example, the City of Chicago police had only five of them in their possession. The police themselves doubtless realised how dangerous the weapon was when used on a crowded street. Nevertheless there were some more bloodthirsty souls who felt that the only way to deal with the hoodlums was to beat them at their own game. As early as 1920, a military writer in *Scientific American* had the following to say about the merits of the Thompson gun:

> The marksmanship of the average policeman with the average policeman's revolver is something to make honest men turn pale, and the women and children duck for the subway . . . As a general rule policemen seem far inferior to yeggs when it comes to hitting things they intend hitting with the pocket gun. I do not know whether the yegg spends some portion of his ill-gotten gains in target practice, or whether his ability to hit, where the officer misses, is due merely to the cussedness of inanimate nature. The fact remains. Wherefore the announcement in the public prints of the adoption by the New York police of the wicked little sub-machine gun is of course interesting to those people who wish to see the customary New York brand of gun play a little less one-sided.
> Without doubt the early future will see the happy coincidence of a policeman skilled in the pointing of the new weapon, and an automobile full of yeggs willing to engage in the customary running gunfight. The result will be the worst-shot-up assortment of crooks that has come to the attention of the coroner.[9]

In 1926, in Chicago, Chief Detective William O'Connor formulated a particularly drastic response to the problem of organised crime. He asked for 500 volunteers from amongst those policemen who had served in France, with the intention of forming them into an *élite* corps. In his first address to these would-be storm-troopers, O'Connor did not pull his punches:

> Men, the war is on. We have got to show that society, and the Police Department, and not a bunch of dirty rats, are running this town. It is the wish of the people of Chicago that you hunt these criminals down and kill them without mercy. Your cars are equipped with machine guns and you will meet the enemies of society on equal terms. See to it that they do not have you pushing up daisies. Shoot first and shoot to kill . . . If you meet a car containing bandits pursue them and fire. When I arrive on the scene my hopes will be fulfilled if you have shot off the top of the car and killed every criminal inside it.[10]

But the use of the Thompson gun was not limited to Chicago, though it was here that it had its hey-day. At least some other locales are worthy of mention. The most bizarre was the southern coast of the United States, around Florida, where, during Prohibition, many owners of small boats ran a flourishing, if risky business by smuggling in liquor. A captain would raise a certain amount of capital, get a crew

'G-man' practising on a Department of Justice firing range

together, and then sail a few miles offshore to meet the main fleet of booze ships. Then he would load up, return to the mainland and sell his cargo. But there was one breed of men for whom all this was a little too much like hard work. They would simply hire a fast boat, intercept the contact boats as they sailed out to meet the fleet, and relieve the captain of his bank-roll. They were known as 'go-through guys', and were the quintessence of the gangster afloat. No matter what the weather, they usually spurned the normal sea-going garb and wore snap-brim fedoras, suits and low-cut shoes. The 'tommy gun' was also *de rigueur*:

> The scene was repeatedly the same; the black-hulled boat with the pluming wake . . . Then the inward sweep and the dark, crouched figures in the last of the sunlight. The Thompson gun was shown, the gunner straddled in the cockpit, his legs braced by a pair of companions so that his aim would be precise.[11]

But the next crime-wave that really caught the public imagination was the outbreak of bank robberies, kidnappings and gun battles that flared across the mid-West of America in the 1930s. To some extent the violence was attributable to the poverty and desperation occasioned by the Depression, and certainly for many poor people living in the area at the time the bandits' exploits made them into Robin Hood figures. They were indeed rather akin to latter-day 'western' heroes. At that time all police forces in America were tied within their state boundaries, and the new-style bank robbers used fast automobiles, instead of the traditional horse, to make their getaway from one state to another. Many now notorious names were involved: John Dillinger, Bonnie Parker, Clyde Barrow, 'Baby Face' Nelson, 'Pretty Boy' Floyd, 'Machine Gun' Kelly and 'Ma' Barker. The story of their exploits is punctuated with the fire of 'tommy guns'.

On 17 June 1933, 'Pretty Boy' Floyd, Adam Richetti and Verne Miller, an expert machine gunner during the First World War, arrived in Kansas City to try and rescue a colleague, Frank Nash, who had recently been arrested. Five policemen were preparing to drive Nash to another jail in Leavenworth when they were ambushed by the three men, all armed with machine guns. In under thirty seconds they succeeded in killing four of the agents, but unfortunately, illustrating one of the inherent defects of the sub-machine gun, they also managed to kill Nash himself. The 'Kansas City Massacre' hit the headlines all over America.

In March 1934, Dillinger linked up with 'Baby Face' Nelson and others and embarked upon a series of bank raids in which machine guns were always the favoured weapon. Even if the raid went off without a hitch, one member or other of the gang could be relied upon to loose off a burst of shots at a ceiling or a plate-glass window. In April the gang found themselves holed up in a lakeside lodge called 'Little Bohemia', but after a long battle with the police they managed to shoot their way out with machine guns.

It might be thought that the exploits of 'Machine Gun' Kelly would also figure largely in a chapter such as this. In fact, he was one of the rather more disappointing of the 1930s hoodlums. The legend grew up later that he owed his violent sobriquet to the fact that he could write the name 'Kelly' on the side of a barn with a machine gun. In fact, for long enough he was not even capable of hitting the side of a barn with that weapon. He had started as a hip-pocket bootlegger in Memphis, and after serving a few months in a penitentiary, moved on to Oklahoma. There he met Kathryn Thorne who decided that she would attempt to turn him into an authentic gangster. She was a little taken aback when Kelly told her that he couldn't abide violence and could hardly distinguish one end of a gun from the other. Nevertheless she brought him a machine gun and started intensive target practice in the Oklahoma countryside. 'Under her careful tutelage he eventually reached a point where he could actually fire the gun without dropping it, and with his eyes open.'[12] Thorne then went around distributing the spent shells from his gun, saying that they were a souvenir from 'Machine Gun' Kelly. But the legend was a little more grandiose than the facts. Shortly afterwards Kelly was caught smuggling liquor into an Indian reservation and was sent to Leavenworth for three years.

Machine guns also figured prominently in the exploits of Bonnie Parker and Clyde Barrow, and in May 1935, when they were ambushed and killed by the police, much of the damage was done with a Browning automatic rifle. The appeal of the 'tommy gun' for the pair comes through in Parker's poetry. One of her verses was called 'The Story of Suicide Sal' and tells the story of a girl who escaped from prison to avenge herself on her accomplice and his lover who had allowed her to take all the blame. The last two stanzas read:

Not long ago I read in a paper
That a gal on the East Side got 'hot',
And when the smoke finally retreated,
Two of the gangsters were found 'on the spot'.

It related the colourful story

Of a jilted 'gangster gal'.
Two days later a 'Sub-gun' ended
The Story of 'Suicide Sal'.

Another poem, 'The Story of Bonnie and Clyde', had been sent to a Dallas newspaper a few days before her death. Towards the end Parker describes her and her partner's plight in these terms:

The road gets dimmer and dimmer,
Sometimes you can hardly see,
Still it's fight, man to man,
And do all you can,
For they know they can never be free.

If they try to act like citizens,
And rent them a nice little flat,
About the third night
They are invited to fight
By a 'sub-gun' rat-tat-tat.

Let me end this brief account of the gangster and the sub-machine gun with an account of two famous shoot-outs, in which some more of the mid-West bandits met their end. On 16 January 1935, 'Ma' Barker and her son Fred were surrounded by Federal agents in a lake cottage in Florida. They refused to surrender and a four-hour gun-battle ensued. At the end of it the police found that both mother and son had been killed. 'Ma' lay on her back, surrounded by spent cartridges, the still hot machine gun clutched in her hands. Dotted around the house the agents actually found several violin cases which had been used, in the prescribed manner, to carry the machine guns into the house.

But the most remarkable death of all was that of 'Baby Face' Nelson in November 1934. He had been spotted and chased by two members of the FBI, and was forced off the road when one of the agents put a bullet in his water tank. The agents, Hollis and Cowley, also stopped and started firing at Nelson from the shelter of a ditch.

Finally Nelson impatiently grabbed a machine gun and said, 'I'm going down there to get those sons of bitches.' A group of men planting trees several hundred feet away were horrified to see 'Baby Face' walk toward the FBI car, upright, his gun blazing from the hip – like a movie gangster. Cowley fired from the ditch as Nelson shot at him. One of Cowley's bullets hit Nelson in the side. But he didn't fall; he kept sweeping the ditch with bullets until Cowley dropped fatally wounded. Hollis was firing his shotgun. Half a dozen slugs tore into Nelson's legs but he plodded forward. Hollis dropped the empty shotgun, pulled out a pistol, and ran for the protection of a telephone pole. Before he reached it a bullet from Nelson's machine gun hit him in the head, killing him.[13]

The key phrase here is 'like a movie gangster'. One wonders, in fact, whether Nelson was not modelling himself upon figures whose exploits he would have had plenty of opportunity to see on the screen. And as Oscar Wilde once said: 'The Americans are certainly great hero-worshippers, and always take heroes from the criminal classes.' The history of the American film industry has certainly done much to substantiate such a point of view. For, from 1930 or thereabouts, the gangster suddenly became a cinematic hero, and his activities passed from sordid reality into celluloid myth. The first gangster film was the silent *Underworld*, made in 1927, but the genre never really got off the ground until the early 1930s and the appearance of films like *Little Caesar* (1930), *Public Enemy* (1931) and *Scarface* (1932). In 1931 alone, fifty gangster films were made, and many of them were great box-office successes. It is often asserted that the bloody happenings of St.Valentine's Day, 1929, started a public revulsion against the activities of the large-scale organised criminals like Capone. Be that as it may, one cannot deny that, at a time when the weekly cinema audience in the United States was 60-65 million, the success of films like *Little Caesar* meant that a sizeable section of the public were not altogether opposed to gangsterism and its values.

These films represented a condoning and even glorification of violence which reflected the cynical state of mind of many Americans at this time, and their belief in the power of force over ideals. It would be going too far to suggest that there was a fundamental restructuring of American attitudes. For at their core there was still a belief in the possibility and the desirability of individual advancement. But there was a new callousness about the means of obtaining it. Thus if the system could not provide it, one was justified in going out and grabbing it for oneself; if the old legal and institutional framework had nothing to offer, then one had to make one's own laws. Though the basic 'American Dream' might still exist, it had gone somehow sour. And the machine gun was the ideal prop for this new version of the American success story. Robert Warshow has pointed out an interesting contrast between the gangster film and the classic western, where the individual is always constrained by a much more orthodox morality: 'The gangster's pre-eminence lies in the suggestion that he may at any moment lose control; his strength is not in being able to shoot faster or straighter than others, but in being more willing to shoot. "Do it first. Do it yourself," says Scarface, expounding his mode of operation, "and keep on doing it." '[14] A burst of machine-gun fire from a moving car is the perfect realisation of such a philosophy. Thus, when the gangster was tranferred to the screen, and his habits and attitudes were shaped into a recognisable iconography, the 'tommy gun' became one of its most recognisable features. With a sub-machine gun blazing from the hip, or from a fast car, the individual was endowed with superhuman powers, literally able to simply sweep any opposition aside. When Colonel Thompson called his gun the 'trench broom' his imagery, if not his choice of locale, was faultless. In this respect, it is interesting to note how many times the St.Valentine's Day Massacre has been portrayed on the screen (e.g. *Scarface*, *Al Capone*, *The Scarface Mob*, *Some Like It Hot*, *The St.Valentine's Day Massacre*). Surely it is not just as a warning against dubious business practice that we have so often seen this ruthless killing re-enacted?

James Cagney on the wrong end of a tommy gun

The pure gangster movie went into decline between 1933

and 1935. The excessively cinematic exploits of the mid-West bandits forced certain public figures and Hollywood moguls to ponder on the social effects of their films. From 1935, FBI agents and various types of *agents provocateurs* became the new heroes, dedicated to eliminating the gangsters. But the 'tommy gun' lived through this transition, becoming the instrument of the gangster's downfall rather than his passport to success. One such film was *Show Them No Mercy* (1937), 'a denunciation of kidnappers, wherein a girl, in heroic and exemplary style, kills the ruthless gangsters with a machine gun.' Since then the gangster movie has known its ups and downs, but Hollywood has always been ready to churn them out, given the slightest encouragement, and whenever they appear the 'tommy gun' is inevitably present (e.g. *The Roaring Twenties* (1940), *White Heat* (1949), *Gun Crazy* (1950), *The Mob* (1951), *Kansas City Confidential* (1952), *Chicago Syndicate* (1955), *Baby Face Nelson* (1957), *Machine Gun Kelly* (1958), *The Scarface Mob* (1959), *The Rise and Fall of Legs Diamond* (1960)).

Recently the gangster film has had yet another revival, prompted by the phenomenal box-office success of *Bonnie and Clyde* (1967). The machine gun once again figured prominently, both in the hands of the bandits as they blasted their way out of an ambush, and in those of the authorities as they

Gary Cooper taking some of them with him in For Whom the Bell Tolls

hunted the gang members down. Indeed, this latter fact reflects a deepening of the cynicism that prompted the original gangster cycle. Though the bandits are indubitably the heroes of the film, unlike the post-1935 cycle mentioned above, they lack the ruthless drive of the very first movie hoodlums. 'Rootless, homeless, anomic, Clyde Barrow is going toward nothing; he is just running from . . . Clyde aspires to no gang leadership: the cut and dried dynamics of the early 'thirties are replaced by a hotch-potch of sociological loose ends.'[16] Thus their own violence is rendered essentially futile as they simply drift, waiting for the forces of law and order to catch up with them. In the latter's hands the machine gun takes on a new significance as the symbol of a ruthless and all-powerful establishment with the firepower to contemptuously eliminate anyone who tries to stand up to it. The final images of Bonnie and Clyde, their bodies jerking like marionettes as the bullets hit them, portrays this notion exactly. Even in death, society can still make us jump.

The machine gun has figured in two other recent, very successful films, and in these also it has been used to put across this notion of the individual's impotence in the face of modern society. In *If* . . . (1968) three public schoolboys and a girl rebel against the hypocrisy and conformity of their education. In the last scene they take to the rooftops with Bren and Sten guns and open up on the parents and teachers emerging from a Speech Day ceremony. But by his non-realistic, almost pantomime portrayal of the carnage that follows, Lindsay Anderson, the director, makes it clear that such an act is at best a fantasy, at worst a futile gesture. As he himself has said: 'Far indeed from filling me with dread, I find the last sequence of the film exhilarating, funny (its violence is so plainly metaphorical) . . . and finally sad. It doesn't look to me as though Mick can win. The world rallies as it always will, and brings its overwhelming firepower to bear on the man who says no.'[17]

In Sam Peckinpah's *The Wild Bunch* (1968), there is also a climactic machine-gun massacre, though this time presented in a vividly realistic manner. A group of outlaws are on the run after staging a bank robbery. They are being relentlessly pursued and are tired and disillusioned, aware that the West they once knew is rapidly changing. There is no longer room for the individual free-booters. They arrive at the camp of a corrupt Mexican warlord, who brutally executes one of their companions. They resolve to stay and fight it out with his men. The disproportionate numbers on each side make it clear that they are in fact resolving to commit suicide, to flee a world they do not understand. They take possession of a Browning medium machine gun and proceed to mow down the warlord's troops until they themselves are killed. In a western, ironically, the nihilism that was always implicit in a gangster film comes completely to the surface. Men find no more meaning in life beyond the act of shooting down as many fellow-beings as possible. They subordinate themselves completely to their gun, and sublimate their helplessness in its deadly efficiency. The machine gun, in fact, becomes the hero of the film. There is no more room in the world for human heroes, and the machines of death have taken over.

In many ways, then, the machine gun had become something of a contemporary icon. The sheer violence of its action, and the indiscriminate deadliness of its effect, has made it a useful symbol for expressing modern man's frenzied attempts to assert himself in an increasingly complex and depersonalised world. One saw at the end of the last chapter that in the First World War the machine gun itself actually helped to engender this feeling of individual irrelevance in the face of the new technology of death. Since then, however, technological innovations have left the machine gun far behind. The machine gun has now become personalised, itself the means by which men desperately try to make their mark on a world in which they feel increasingly powerless. In the fantasy world at least technology is turned against itself.

Notes

1. W.J.Helmer, *The Gun That Made the Twenties Roar*, Macmillan, New York, 1970, pp.72 and 78.
2. Captain E.C.Crossman, 'A Pocket Machine Gun', *Scientific American*, 16 October 1920, p.413.
3. Helmer, op.cit., p.77.
4. J.Kobler, *Capone*, Coronet Books, London, 1973, p.91.
5. F.Watson, *A Century of Gunmen*, Nicholson and Watson, London, 1931, p.185.
6. Kobler, op.cit., p.177.
7. K.Allsop, *The Bootleggers*, Four Square Books, London, 1963, p.285.
8. J.H.Lyle, *The Dry and Lawless Years*, Prentice-Hall, Englewood-Cliffs, 1961, pp.82-3.
9. Crossman, op.cit., p.405.
10. Allsop, op.cit., p.128.
11. R.Carse, *Rum Row*, Jarrolds, London, 1961, p.127.
12. M.J.Quimby, *The Devil's Emissaries*, A.S.Barnes, New York, 1969, p.21.
13. J.Toland, *The Dillinger Days*, Mayflower, London, 1965, p.270.

14. R.Warshow, 'The Westerner', in D.Talbot (ed.), *Film: an Anthology*, University of California Press, Los Angeles, 1969, p.152.
15. L.Jacobs, *The Rise of the American Film*, Harcourt Price, New York, 1939, p.513.
16. A.Bergman, *We're in the Money: Depression America and Its Films*, New York University Press, New York, p.171.
17. L.Anderson and D.Sherwin, *If . . .*, Lorrimer, London, 1969, pp.12-13.

VII *New Ways of War*

Obviously the history of the machine gun as a military weapon goes on beyond 1918. No army in the world would now dream of not having its complement of medium, light and sub-machine guns. Nevertheless, the weapon's history since the end of the First World War has become increasingly less useful as a central focus upon general civilian and military attitudes. For one thing such guns were not now the unknown quantity they had been at the beginning of the war. Their awful effectiveness had been seen too often to be ignored; whilst the creation of *élite*, specialist bodies such as the Machine Gun Corps had permitted the formulation of a detailed set of machine-gun tactics. By 1918 the Corps staff had worked out complicated tables and equations to cover all aspects of the battlefield use of their Vickers guns. Moreover, both sides had also begun to come to terms with light machine guns, using them as the key mobile weapon for small groups of attacking infantry.

But it was not simply that the machine gun was more accepted and better understood. It had now also lost its decisive importance as an obstacle to attacking infantry. For the First World War had thrown up another weapon that meant the end, sooner or later, of the machine gun's battlefield dominance. The very recognition of this dominance had meant that some men at least immediately set to work to try and find a means of nullifying it. Their solution was the tank, an armoured vehicle meant to obviate the necessity for infantrymen to cross open ground in front of the enemy's gunners. The vehicles would either carry the infantry themselves, or they would be used to smash a hole in the enemy's defences, principally by destroying his machine guns, through which the infantry could then pass.

The contemporary records make it clear that the tank was first and foremost a response to the power of automatic fire. The Landships Committee, in 1915, observed that: 'These landships were at first designed to transport a trench-taking storming party of fifty men with machine guns and ammunition.' In a memorandum of the same year, Swinton, one of their most fervent advocates, wrote: 'Wherever it has been possible beforehand to mark down machine gun emplacements in the German front line the destroyers will be steered straight at them, will climb over them and will crush them. At other points they climb the enemy's parapet or trench and halting there will fire at any machine guns located with the two-pounder gun . . .'

In the years that followed theories about the role of armour became more sophisticated and it soon transcended its function as a mere battering ram. Its very existence rendered pointless the construction of trench lines like those on the Western Front. Under the guidance of theorists like Fuller, de Gaulle and Guderian, the tank became one of the principal weapons around which quite new types of land tactics were evolved. The very existence of the tank, even the crude models of the 'twenties and 'thirties, meant that the power of the static defence had been considerably weakened. Though there were still many who refused to acknowledge this fact, it was finally demonstrated with a decisiveness to convince all but the most obtuse in 1940, when the German *Blitzkrieg* tactics were pitted against the French Maginot Line. Armoured mobility was now the decisive factor in war and the relative importance of the machine gun declined accordingly. As anyone who has read Liddel-Hart's *The Tanks* will appreciate, this in no way means that the various military establishments adopted tanks without a murmur. The same kind of conservatism and myopia that had dogged the development of the machine gun also continually thwarted the best efforts of the proponents of armoured warfare. But what is important in terms of this book is that the conservatives' *point d'appui* had changed. The role of the machine gun had ceased to be a crucial issue.

Certainly it was still an important weapon. Medium machine guns could still be used to great effect in repelling infantry assaults*, consolidating captured positions or cover-

---

* As in Korea, for example, against the Chinese 'human wave' assaults.

ing a withdrawal. Light guns were indispensable for laying down a localised fire-field under cover of which the platoon and the squad could make their sporadic advance. But the machine gun was no longer decisive. It could not now, on its own, affect the outcome of a large-scale offensive. On the Western Front it had been the machine gun more than anything else that had destroyed the attackers at Verdun, Champagne or Passchendaele. Because of it unprotected infantry were incapable of crossing the gap between the trench lines in sufficient numbers to punch a gap through them. So one after another the great offensives ground to a bloody halt and the trench lines remained as they were. After 1918 the development of the tank and other armoured vehicles, coupled with the elaboration of more sophisticated infantry tactics, denied the machine gun the opportunity to create such wholesale havoc. Now men were shielded by armour plate and mobile guns. Even if they were in the open they no longer offered the enemy such targets as the parade ground formations that tried to stroll across no man's land. Men were now aware of their powerlessness in the face of such mechanical weapons. The machine had finally earned a little grudging respect and gone were the days when whole battalions might be mown down by just one machine gun, or a whole colony might be at risk just because the Maxim jammed. Men had entered a new military era.

War Without Machines

But the transition was far from easy. No social group was able to adapt effortlessly to the new demands of an industrialised society, and the military establishments found the process particularly painful. Men are always slow to adapt their ways of thinking, to modify their old assumptions in the light of new economic and technological developments. The

*Machine guns in
People's Square
Shanghai 1963*

169

laggardliness of nineteenth-century armies in this respect is a particularly striking example of the resilience of out-dated modes of thought. And a study of their attitudes to the machine gun offers a particularly good way of highlighting the various social and political factors that were of importance during this particular watershed in European thought. On the other hand it reveals just how closely the development of military theory and practice is tied up with the nature of society in general, both the concrete social and economic conditions and men's way of perceiving them. Military history, like everything else, is a social phenomenon, and even the weapons themselves have their social history.

For hundreds of years the notion of automatic fire was just not a viable proposition. I pointed out earlier that one of the reasons for the general lack of interest were the almost insurmountable technical problems: though some of these ideas were conceptually sound they were always doomed never to get beyond the drawing board. For right up to the middle of the nineteenth century men simply did not have the techniques or the materials to translate such ideas into reality. Throughout the preceding centuries technology, and military technology in particular, remained almost static. But such a bald statement rather begs the question. For the level of technical expertise itself did not exist in an intellectual vacuum. Inventiveness had to be stimulated by profound economic and social changes. In the last analysis it was the absence of such changes, the static nature of society as a whole, that determined early reactions to the machine

Chinese gansters executed in Shanghai 1949

gun. The technical base for machine gun production did not exist because men did not see the need for them.

For centuries the states of Europe were all dominated by a land-owning *élite*. They obtained sufficient revenue simply from the rents they charged their serfs or their tenants. Thus they had very little interest in encouraging any kind of mechanical innovation. There was of course in all countries an important non-landowning middle-class element, but they almost all made their money by dealing in raw materials or by providing financial and legal services. They too had little interest in technological progress. Also, the actual production of consumer goods that was undertaken was done in very small establishments, generally an artisan's workshop, where the tiny size of the work force made any mechanisation only marginally significant. On top of this, the small numbers of people in any country who could actually afford such goods never justified any attempt to increase output significantly. And if the mode of production does not change then there will be no particular stimulus to technological innovation. Thus technical skills and knowledge remained at a rudimentary level, the effects of this being felt in the realm of military technology as much as any other.

But it was not simply a case of military men being forced against their will to make do with unsatisfactory weapons. For the same lack of expertise that kept gun-making on a crude level had other more basic effects that kept most military men quite happy with the firearms they had. Experimentation in the field of automatic fire was further hampered by the fact that most influential soldiers did not really see the need for it.

The states of Europe before the Industrial Revolution were always in financial difficulties, and even at the best of times governments had only a very limited amount of money to spend on the military. Thus they could not afford to train, pay or equip particularly large armies. More importantly, they could not afford to raise or maintain reserves with which to replace heavy casualties on the battlefield. Secondly, even if the states had been able to afford such supplementary military establishments, they did not have the administrative machinery to mobilise them. Finally, horse-drawn transport and very poor roads made it impossible for any government to supply or effectively concentrate anything other than a fairly small army.

For all these reasons, then, states regarded the individual soldier as a very precious commodity. At one point in his writings Frederick the Great flatly stated that: 'One of the most essential duties of generals commanding armies or detachments is to prevent desertion.' He then went on to give a detailed list of fourteen rules that must be followed in this respect. Certainly soldiers at this time were brutalised by vicious methods of discipline, and drilled until they were little more than automatons. But this did not detract from the fact that the state was most anxious to keep them alive. For the army in the field was literally irreplaceable, at least in the short term. No state had the financial or the administrative capacity to emulate those of a later century, for example Britain in 1915, who could throw whole new armies into the field once the original force, in this case the BEF, had been virtually destroyed. Therefore no government or military establishment saw much advantage in developing weapons that could destroy men quicker than they could be replaced. To most men before the Industrial Revolution, and beyond, the notion of greatly improved fire-power simply did not make much sense. So those few visionaries who sporadically threw up the concept of automatic weapons found themselves caught in a vicious circle. Society had not yet developed a technical base adequate to the manufacture of such weapons, whilst the very backwardness of states' administrative and financial structures rendered even the idea of such a massive increase in firepower essentially futile. The world before the middle of the nineteenth century was simply not ready for the machine gun, and its earliest proponents are now nothing more than historical footnotes.

Machines and Mass Society

By the beginning of the nineteenth century it was possible to distinguish the emergence of a new type of society. For the first time land was beginning to lose its absolute primacy in economic affairs. The middle classes were beginning to emerge as a political force in their own right. Their increasing power and wealth enabled them to think of investing considerable amounts of money in non-agricultural production. At the same time the population of Europe was increasing and this forced impoverished sections of the peasantry into the towns where they swelled the labour force for an expanded production. As the proportion of the population in the towns grew ever larger it created a large, uniform, easily accessible market for cheap, standardised household items. As this market grew it enabled production to be concentrated into larger factories whose owners, assured a large enough demand, could afford to attempt economies of scale and mass production methods. As this network of factories grew, and the market became more homogeneous, so did the demand for heavy machinery and the wherewithal to create an efficient communications system, notably a railway network. This demand stimulated the expansion of the iron and steel and heavy engineering industries and drew even greater numbers of people into the cities. It is difficult to place all these steps in precise sequence, but all that is important here is the realisation that, taken together, over a period of some decades, they added up to what we know as the Industrial Revolution.

Such massive changes in the organisation of society inevitably had major implications for the conduct of war. Naturally enough they stimulated great technological advances as men struggled to produce the machines – and the machines to make machines – that could keep pace with the ever-increasing volume of production. The most vital offshoot of the development of productive techniques was the appearance of machine tools, complex lathes, drills, files etc. which could be relied upon to do the same job, in exactly the same way, time and time again. In this way the production of the finished item, be it a needle or a steel plate, a button or a cannon, was taken out of the hands of the individual craftsman and done on the infinitely faster and more reliable production line. In the field of armaments this meant that individual parts could be made to much smaller tolerances, so that each component fitted together exactly. At last it could be guaranteed that what was depicted on the drawing board would look just the same in reality, and this would be so for every article that was produced. Moreover, the great strides made in the field of metallurgy meant that not only would the individual components fit together and work together as they should, but also that they would be strong enough to withstand the strains of frequent and prolonged repetitive cycles. Such complex items as the machine gun were now a quite feasible technical proposition.

But the development of these industrial techniques had more than mere technical implications. The new productive potential brought with it other far-reaching changes that gave an impetus to the development of automatic fire. It had, for example, a significant impact on the conduct of war. This was first felt in America during the Civil War. There it was found that the techniques of mass production, and the enormous improvements in communications that had accompanied the expansion of the market, meant that each side in the war was able to field very large armies, and to raise and equip huge numbers of reserves to replace the casualties. The Americans found that industrialisation and mass production had brought with them mass war, in which an unprecedentedly large proportion of the population could be funnelled to the front. It would seem to be no coincidence that it was in America, the first country to experience this new type of warfare, that the first workable machine guns appeared. For the machine is above all else the weapon of mass warfare, the ideal arm for a conflict in which the indivual soldier is expendable. In the War between the States warfare showed the first signs of the transition to total warfare, in which nations pitted their whole population and their whole productive capacity against each other. In such wars it is essential to kill as many of the enemy as possible in the quickest, most economical way. For this the machine gun was ideal.

There is one further reason why workable automatic weapons should have appeared at this time, as a spin-off from the progress of industrialisation. For this process had brought with it a whole new ideology. Now that the floodgates of industrial capitalism had been opened, those riding on the crests of its waves saw the beginning of a new era. They believed quite sincerely that the new potential of these vast increases in productivity meant the approach of an undreamt of prosperity. Machines would be the new saviours of humanity, creating wealth and material goods for all. This belief in the beneficient effects of technology was very much apparent in the field of armaments. It has already been noted how Gatling thought that automatic weapons would actually *reduce* casualties. An obscure British writer summed up this blind optimisim as well as anybody. He asserted, in a childrens' book, that: 'The man who invents the most rapid and the most effectual means of destruction, as regards war, is the greatest friend to the interests of humanity . . . If any man could invent a means of destruction, by which two nations going to war with each other would see large armies destroyed . . . in a single campaign they would both hesitate at entering upon another . . . In this sense the greatest destroyer is the greatest philanthropist.'[1] In a century in which such optimism was commonplace it is not to be wondered at that many men turned their minds to the perfection of machines that could kill their fellow men as efficiently as possible.

The Military Mind

But the mere appearance of the machine gun did not imply its wholehearted endorsement by the relevant men of influence. Certainly it is fair to say that industrial capitalism was becoming the dominant mode of production, and as such it threw up a whole school of thought eager to preach its virtues. But not everyone shared this burning faith in the potential of factories and machines. For some groups these very machines were a threat, a symbol of the new world in which their old dominance and self-confidence was being undermined.

One such group was the nobility and gentry of the various European nations. They had been most powerful in a world based upon agricultural production and the ownership of land. The sudden transition to an economy based around agricultural production and the city threw up a whole new group of potential leaders with a quite different set of values. But the old landowners were not thus completely bereft of influence. One of their traditional strongholds had always been the army, and there they remained in force right up until the outbreak of the First World War. Within the army, throught the nineteenth century, they continued to express their distrust of all the products of the new industrial society.

They refused, almost to a man, that the emergence of that society had made hardly any difference to the way that war was to be conducted. Ever since the days of Frederick the Great the military establishments of Europe, with the temporary exception of France during the First Republic, had known a remarkable continuity, with fathers, sons and grandsons passing through the same regiments, and absorbing and disseminating the same orthodoxy about the unchanging nature of war. Even worse, the rigid hierarchical organisation of the army made it into a gerontocracy, which ruthlessly discriminated against any initiative or originality from below. The easiest way to promotion lay in due deference to one's elders and an unquestioning acceptance of the *status quo*.

Thus, though the machine gun, *qua* machine, can be regarded as an inevitable product of the nineteenth-century faith in the benefits of science and technology, its inventors soon came up against a quite contradictory set of values amongst the military. The latter's conception of warfare was firmly rooted in the past, in an age when the musket, bayonet and horse, particularly the latter, had been the decisive weapons on the battlefield. In this type of warfare the individual officer and gentleman still felt that he counted for something. It was still possible to conceive of situations in which men alone, if possessed of sufficient courage or an adequate sense of honour, could make a decisive mark and turn the course of events. Such attitudes were not only kept alive by the inbuilt conservatism of the military establishments, they were positively reinforced by the officers' desire to reassert their individuality in a world in which their traditional social dominance was fast being eroded.

The machine gun represented the very antithesis of this desperate faith in individual endeavour and courage. One had only to see the thing being demonstrated to realise that the force of such a deluge of fire, directed by just one man, could sweep away whole units with nonchalant ease. Neither the resolution of the opposing gunner nor of the troops in his sights would count for anything. The weapon itself, a mere machine, completely dominated the situation. As long as it did not jam or run out of ammunition the men themselves were helpless before it. Cavalry or infantry, officer or soldier, coward or hero, all could be bowled over like rats before a hosepipe. So the inventors and their agents, inordinately proud of having introduced the benefits of technological progress into war itself, were chagrined to find that the military simply did not want to know. In the machine gun it dimly perceived a basic threat to the old certainties of the battlefield and, with a remarkable degree of success, it doggedly refused to admit the machine gun's very existence.

Even the success of the weapon in far-flung parts of the world had little impact on the military mind. What one might call ideological factors are again of importance here. For these terrible tests of the efficacy of automatic fire were carried out against 'savages' in Africa, India or Tibet. No matter how much one tries to doll up imperialism with high-sounding rhetoric about 'the white man's burden' or the 'civilising mission', it is undeniable that racism was one of the central features of this colonial expansion. To be able to sustain their ruthless drive for new markets the European nations had to believe that they were inherently superior to the hapless natives. In this way it was possible to justify their wholesale slaughter or expulsion from their traditional lands. But this contempt for the indigenous populations meant that most soldiers did not regard colonial campaigns as being warfare proper. The fate of a horde of natives before the immense firepower of a small European expeditionary force was not deemed in the least analogous to what might happen on a future European battlefield. Mowing down Matabeles, Dervishes or Tibetans was regarded more as a rather risky kind of 'shoot' than a true military operation. So the soldiers were able to blithely ignore the obvious lessons about the devastating potential of automatic weapons. Ignorant natives might try and rush stockaded machine gun positions but civilised European armies were surely well above such tactical absurdities. The machine gun's firepower could not possibly be relevant in such a bloodily crude fashion.

On top of this, because of the army's own self-esteem and the fierce jingoistic pride of the nation as a whole, it simply would not do to admit that their successes against the overwhelming odds, on paper at least, that they faced in Africa was a result of superior weaponry. Contemporary readers

The Gatling gun lives on after one hundred twenty-five years. After World War II, the General Electric Company undertook development of a modernized version of the Gatling that would deliver high rates of fire from aircraft, ground vehicles, and ships.

Dr. Gatling's gun still is an awesome source of firepower. Here a U.S. Army technician fires a 20mm M61 Vulcan automatic cannon built by the General Electric Company. Its rate of fire is 6,000 shots per minute. This modernized Gatling is one of several models mounted on contemporary American fighter aircraft.

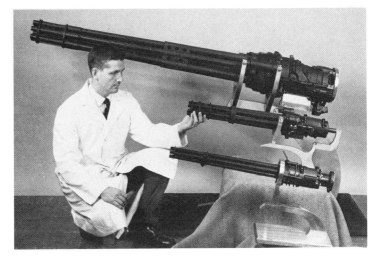

The twentieth-century progeny of Dr. Gatling's hand-cranked machine guns. From top to bottom: the 20mm Vulcan aircraft gun and the 7.62mm and 5.56mm Miniguns. All three of these General Electric machine guns are externally powered by electric motors and can be fired at rates as high as 6,000 shots per minute.

Since World War II, development of the machine gun has continued. One of the leading post-war guns, the Fabrique Nationale (Belgium) *Mitrailleuse d'Appui General* (MAG) is clearly the most popular, having been adopted by three times as many countries as its nearest competitor, the Rhein-metall MG3. Of the seventy or more countries that have adopted the MAG, Belgium, Sweden, the United Kingdom, Israel, India, the United States, Nigeria, and Argentina also manufacture this GPMG for their armed forces. Approximately 150,000 MAGs have been manufactured to date by Fabrique Nationale and its licensees.

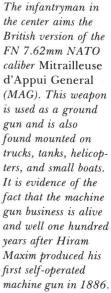

The infantryman in the center aims the British version of the FN 7.62mm NATO caliber Mitrailleuse d'Appui General (MAG). This weapon is used as a ground gun and is also found mounted on trucks, tanks, helicopters, and small boats. It is evidence of the fact that the machine gun business is alive and well one hundred years after Hiram Maxim produced his first self-operated machine gun in 1886.

Monsieur Ernest Vervier holds a production version of the 7.62mm NATO caliber Mitrailleuse d'Appui General (MAG), which he evolved from the Model 1918 Browning Automatic Rifle created by the American design genius John M. Browning.

178

wanted to hear that it was the heroism and innate superiority of the Englishman that had carried the day, not a machine that could fire hundreds of bullets per minute. So the role of firepower was played down, almost ignored. With very few exceptions, it is in vain that one scans the pages of contemporary journals such as the *Illustrated London News* or the *Graphic* for any pictures of machine guns actually in action against the natives. Like everyone else at that time the illustrators preferred to dwell upon the thin red line or square of doughty heroes. Nobody wished to know about the real reasons for their success.

The Changing Shape of War

We have, then, a basic contradiction. On the one hand there is the progress of industrialisation and the proliferation of the machine. On the other is the innate conservatism of the military establishment and the general belief that an army was the sum of its human parts rather than mere operatives of its weapons. Eventually the contradiction had to resolve itself. This is what happened on the Western Front between 1914 and 1918. Almost unknowingly, by 1914, armies had come to depend on weaponry – artillery, rifles and machine guns – rather than upon the soldiers themselves. This became quite apparent in the first months of the war, as the soldiers desperately dug holes in the ground to escape from the firepower they barely realised had been created. The generals never came to terms with this power. Time and time again they threw their men forward, confident that this time a little more preparation, a few more men, and an extra dash of sheer courage would suffice to break the enemy's will to resist. They never realised that they were not fighting his 'will', but his machine guns. And they were implacable and unshakeable. Morale was an irrelevancy to them; all they needed was enough water and bullets. The man counted for nothing. The machine had taken over. Thus Guy Chapman's poignant eulogy to those going up to the front on the Somme: 'Hump your pack and get a move on. The next hour, man, will bring you three miles nearer to your death. Your life and your death are nothing to these fields – nothing, no more than it is to the man planning the attack at GHQ. You are not even a pawn. Your death will not prevent future wars, will not make the world safe for your children. Your death means no more than if you had died in your bed . . .'[2]

By 1918 the transition from the old type of war to the new was complete. The machine gun had played a major part in this transition, highlighting the contradictions and, eventually, hastening the passing of the outdated beliefs about the importance of the individual. In response to the machine gun the tank was developed and mechanised warfare advanced one stage further. Increasingly the quality of a country's weaponry and the capacity of its industrial output became the determinants of success, rather than any will to win born of idealism, faith or personal self-respect. This dehumanisation of war has continued unabated. On the conventional battlefield men are increasingly being replaced by electronic devices. In the mechanised era men had at least to drive the tanks, aim the guns and pull the triggers. Now such things are done by computers and infra-red homing devices. Men are merely helpless bystanders. With the advent of nuclear weapons this process has been carried to its logical conclusion. Now the destruction of the whole world is contingent upon the mere pressing of a button. And we might yet suffer the supreme irony that even the personality of the button-pusher will not intervene between survival and annihilation. Nuclear weapons are controlled by unimaginably complex electronic systems. It is almost statistically inevitable that one day one of these systems will malfunction, and its complexity will make it almost impossible to trace the fault in time. By then the missiles will have left the silos. The world will go neither a whimper nor a bang, just a simple short-circuit.

This book has been the history of man's relationship with his technology. One of its purposes has been to discover how well he coped with the necessity to re-examine his old modes of thought in the light of the terrible potential of automatic weapons. It seems clear that on the whole he coped very badly indeed. Today we live in the shadow not only of nuclear holocaust but also of total pollution or the exhaustion of the world's resources. The technology that was designed to comfort us or to make us feel more secure now seems to threaten us from every side. There are those who argue that technology can yet save man, but those optimists who foresee him being conveyed triumphantly into a neon sunset would do well to ponder the history of the machine gun. Even this brief study has revealed numerous examples of man's greed, hypocrisy, blindness and callousness. Perhaps we shall be able to approach the problems that face us today with a clearer and more honest outlook. But the history of the machine gun should at least teach us something about man's potential for self-destruction. Let us hope that the whole story will not be simply a footnote to Armageddon.

Notes

1. Clarke, op.cit., p.77.
2. Chapman, op.cit., p.95.

Bibliography

Many books have been consulted in the course of writing this book. Those that have proved particularly useful have been mentioned in the notes to each chapter. In order to avoid undue repetition this bibliography only cites those works that deal exclusively with the machine gun or with firearms in general.

Allen G.W.B., *Pistols, Rifles and Machine-guns*, English Universities Press, London, 1953.

Browning J. and Gentry C., *John M. Browning: American Gunmaker*, Doubleday, New York, 1964.

Chinn Lt.-Col. G.M., *The Machine-Gun*, Bureau of Ordnance, Department of the Navy, Washington, 1951–53 (4 vols.).

Ffoulkes C., *Arms and Armament*, Harrap, London, 1945.

Hobart F.W.A., *Pictorial History of the Machine-Gun*, Ian Allen, Shepperton, 1971.

Helmer W.J., *The Gun that Made the Twenties Roar*, Macmillan, New York, 1970.

Hobart F.W.A., *Pictorial History of the Machine-Gun*, Ian Allen, Shepperton, 1971.

Hoff A., *Airguns and Other PneumatArms*, Barrie and Jenkins, London, 1972.

Holley I.B., *Ideas and Weapons*, Yale University Press, Yale, 1953.

Hutchison Lt.-Col. G.S., *Machine-Guns: their History and Tactical Employment*, Macmillan, London, 1938.

Lake W.R., *Machine-Guns and Automatic Breech Mechanisms*, Haseltine Lake, London, 1896.

Longstaff F.V. and Atteridge A.H., *The Book of the Machine-Gun*, Hugh Rees, London, 1917. (This book also contains a complete bibliography of machine-gun literature for the years 1860–1915.)

Pridham C.H.B., *Superiority of Fire*, Hutchinsons, London, 1945.

Wahl P. and Toppel D.R., *The Gatling Gun*, Herbert Jenkins, London, 1966.

Bibliographical Essay

The Social History of the Machine Gun is unique in that it is a discussion of the social evolution of machine gun technology and usage. Nearly all of the other histories of machine guns concentrate on the evolution of these weapons simply as elements of mechanical technology. Ellis has indicated his basic source materials at the end of each chapter and in his bibliography. Listed below are some of the more important books relating to the historical development of the machine gun.

Although flawed, Lt. Col. George M. Chinn's four-volume study, *The Machine-Gun* (Washington, D.C.: Bureau of Naval Ordnance, Department of the Navy, 1951–53), is the essential starting point for all studies of automatic weapons. The first volume summarizes the history of machine guns to the end of World War II. Volume four presents technical design data, schematic drawings of weapon mechanisms, and patent and bibliographical information. The formerly classified volumes two and three deal respectively with Soviet automatic weapons history and the development of machine guns and automatic cannon during the years 1941 to 1950.

Prior to the publication of Chinn's volumes, two British officers, Maj. F. V. Longstaff and Capt. A. Hilliard Atteridge, had compiled much useful data in their seminal study, *The Book of the Machine Gun* (London: Hugh Rees, 1917). Longstaff and Atteridge introduce their book by saying: "The soldier is by nature conservative. Mature age and high command usually go together, and it is the exceptional man who, as years increase, maintains the openness and elasticity of mind that welcome new ideas. Among the younger men, until very recent times, an officer hardly improved his prospects by being a seeker after novelties. It was a sounder policy to accept the existing regulations and the traditional methods as sufficiently near to perfection for all practical purposes." They then proceed to elaborate upon the relationship of military conservatism to the adoption of the machine gun. *The Book of the Machine Gun* is also useful for its very complete bibliography of literature on machine guns for the period 1860 to 1915.

Equally important for the early period is W. R. Lake's *Machine-Guns and Automatic Breech Mechanisms* (London: Haseltine Lake, 1896).

Lt. Col. G. S. Hutchison recorded the reception of the

machine gun in his book *Machine Guns: Their History and Tactical Employment (Being also a History of the Machine Gun Corps, 1916–1922)* (London: Macmillan and Co., 1938), which was published on the eve of World War II. G. W. B. Allen, *Pistols, Rifles and Machine-guns* (London: English Universities Press, 1953), brings the story of machine gun development up to the conclusion of World War II.

On the American scene, Capt. Julian S. Hatcher, 1st Lt. Glenn P. Wilhelm, and 1st Lt. Harry J. Malony tried to educate the United States military with their book *Machine Guns* (Menasha, Wis.: Collegiate Press, George Banta Publishing Co., 1917).

Equally informative from a historical vantage point are: Paul Wahl and Donald R. Toppel describing the evolution of *The Gatling Gun* (New York: Arco Publishing Co., 1965); John Browning and Curt Gentry, *John M. Browning: American Gunmaker* (New York: Doubleday, 1964); and William J. Helmer's study of the Thompson submachine gun, *The Gun That Made the Twenties Roar* (New York: Macmillan Co., 1970).

Should a reader desire to pursue in more detail the study begun by Ellis of American machine gun development and the relationship of that technology to the military bureaucracy, one could start with volumes such as David A. Armstrong's *Bullets and Bureaucrats: The Machine Gun and the United States Army, 1861–1916* (Westport, Conn.: Greenwood Press, 1982).

Other recent books on machine guns include: Daniel D. Musgrave and Smith Hempstone Oliver, *German Machineguns* (Washington, D.C.: MOR Associates, 1971); Thomas B. Nelson and Hans B. Lockhoven, *The World's Submachine Guns [Machine pistols].* Vol. I (Cologne: International Small Arms Publishers, 1963); Thomas B. Nelson and Daniel D. Musgrave, *The World's Machine Pistols and Submachine Guns, 1964–1980.* Vol. IIA (Alexandria, Va.: T. B. N. Enterprises, 1980); Daniel D. Musgrave and Thomas B. Nelson, *The World's Assault Rifles and Automatic Carbines.* Vol. II (Alexandria, Va.: T. B. N. Enterprises, 1967); and F. W. A. Hobart, *Pictorial History of the Machine Gun* (London: Ian Hamilton, 1971). For the modern era, Edward C. Ezell's *Small Arms of the World: A Basic Manual of Small Arms* (Harrisburg, Pa.: Stackpole Books, 1983) is a useful reference. It presents a historical essay on infantry weapon development, including machine guns since 1945, as well as country by country discussions of weapons currently in use.

Because it discusses the social context of military technology, Ellis's *Social History of the Machine Gun* has become part of a broader body of literature that has appeared in recent years. Premier among these books is I. B. Holley, Jr., *Ideas and Weapons: Exploitation of the Aerial Weapon by the United States during World War I: A Study in the Relationship of Technological Advance, Military Doctrine, and the Development of Weapons* (New Haven: Yale University Press, 1953; and reprinted by the Office of Air Force History, 1985). Holley, after essaying on the structure of the military profession and its social relationships to its technology, looks at the impact of the military airplane on the way in which the military officer lived and at how it affected his social structure. Holley's study can be read as a parallel case study to *The Social History of the Machine Gun.*

Following Ellis's book in time, Daniel R. Headrick's *Tools of Empire: Technology and European Imperialism in the Nineteenth Century* (New York: Oxford University Press, 1981) pursues many similar themes, especially as they relate to other nineteenth-century technologies employed by European powers to master their colonial dominions. In the same general category of books examining the social and political impact of military technology, but far more wide ranging in its scope, is William H. McNeill's *Pursuit of Power: Technology, Armed Force, and Society since* A.D. *1000* (Chicago: University of Chicago Press, 1982), which examines the mechanization of death from the fourteenth century "arrow factories" to our age of nuclear weapons.

Much of the Ellis text discusses the inability of the European and American military organizations to fathom the impact of new weapons on their craft and their social structure—and on their own individual survival. In addition to understanding Holley's theory on this topic, one should juxtapose it against the discussion in Noel Perrin's *Giving Up the Gun: Japan's Reversion to the Sword, 1543–1879* (Boston: David R. Godine, Publisher, 1979). Perrin makes the point that the Japanese samurai quickly realized that allowing the deadly new Western firearms to reach the hands of peasant soldiers would spell the end to the samurai as a class of sword-wielding warriors. Rather than adapt to a new type of warfare, as had the knights of western Europe, the samurai's solution was to ban gunpowder firearms. This ban lasted three centuries, until the military leadership of Meiji Japan decided to join the Western powers in the age of modern industrial armaments. Perrin's book, taken with Ellis's presentation, provides much food for thought about the consequences of developing and assimilating new weapons into a culture.

My study of rifle development in Imperial Russia and the Soviet Union, *The AK47 Story: Evolution of the Kalashnikov Weapons* (Harrisburg, Pa.: Stackpole Books, 1986), describes the industrial/economic and foreign policy, as well as military, consequences of large-scale automatic weapons production for a major power such as the Soviet Union. The conclusion of the text notes:

> One of the consequences of this concurrence of [Soviet] manufacturing capacity and development capability is the almost ubiquitous nature of Kalashnikov's weapons. As noted at the outset, at least 55 nations use the AK47, AKM, and the Kalashnikov machine guns on a regular basis. These weapons are seen in the hands of little children in Lebanon and Afghanistan. They are used by government and antigovernment forces from Indochina to Central America and Africa. . . .
>
> It should be no wonder then that Mikhail Timofeyevich Kalashnikov is a social and economic hero in his native land. In a land that honors its technological elite, Kalashnikov stands out because his family of successful weapons have given reliable firepower to his nation's infantrymen, and full employment to his weapons factory's employees. And beyond that the name Kalashnikov is known world-wide. In backwater regions, where a Russian has never been seen, men and women equipped with AKs know that Russians make good reliable weapons. It was once noted only semi-in-jest that Americans export Coke, the Japanese export Sonys, and the Soviets export Kalashnikovs.

Discussing a much earlier period than that covered by Ellis's book, David Ayalon's *Gunpowder and Firearms in the Mamluk Kingdom: A Challenge to a Medieval Society* (second edition, London: Frank Cass and Co., 1978) is also concerned with the social impact of new military technology. In this case study of events in western Asia and Egypt at the end of the fifteenth century, Ayalon looks at the manner in which firearms affected a military aristocracy of cavalrymen, and he presents an "analysis of the clash between the deep rooted antagonism of a military ruling class of horsemen to firearms, on the one hand, and the steadily growing, . . . inescapable necessity of employing them, on the other." This volume will provide students a useful contrast and comparison with the books by Ellis, Holley, and Perrin.

E.C.E.

Index

Africa, imperialist wars against Ugandans, 88 against Zulus, 82-4; in Egypt, 85-7, Northern Nigeria, 96-7, Rhodesia, 89-90, Tanganyika, 88-9, Transvaal, 91; definition of European rule in, 92; German East Africa, 92; nationalism in, 100-1

African troops, 95

Africans, their use of machine guns, 95-6

Ager gun, 25

Allenby, General, 59

American Army, ignore firearms developments, 74; reject Lewis gun, 74-6; size of, 71

American Civil War, 21, 23-5, 47, 51, 62-3 anti-Imperialism, 99-100

Anzac Corps, 139

Archduke William (Austrian), 36

Aristocracy, 48-9, 175; as supporters of machine gun, 57

Armies, before Industrial Revolution, 171-2

Armoured vehicles, 168-9

Arms manufacturers, growth in USA, 21-4; in 19th century, 14-15; naive attitudes of, 174; profits of, 39-42

Army and Navy Gazette, 84, 102-3

Artillery, attitude to machine gun, 55, 63-5

Ashantis, expeditions against, 82-4; wars against, 94-5

Auto-Ordnance Company, 149

Baker-Carr, Brigadier-General, 118, 122-3

Baldwin-Felts Detective Agency, 44

Barker, 'Ma', 159

Barnett, Corelli, 106-7

Bayonet fighting, training for, 126-8

Belloc, Hilaire, 94

Benet-Mercie machine gun, 74

Bessemer, Sir Henry, 16

Block, Ivan (Polish banker), 52

Blunden, Edmund, 137-8

Boers, 91

Bonnie and Clyde (film, 1967), 162-3

British, hostility to machine guns, 62-3

British Army, adopt Gatling gun, 63-82; at battle of the Somme, 132-9; brutalization of, 142-5; drill in, 53; faith in cavalry, 54-6, 128-30; ignore machine guns before First World War, 68-9; in offensive at battle of Neuve Chapelle, 131-2; outmoded training of, 124-8; reliance on infantry charge, 133

British Expeditionary Force, 69, 120

British Standard Arms Company (BSA), 35, 40

Browning, William, 16, 38, 40

Browning gun, 37-8, 40

Cabin Creek (USA), 43-4

Campbell, Colonel (bayonet-fighting expert), 126-8

Capitalism, rise of, 172-4; spirit of seen in machine gun development, 15

Capone, Al, adopts sub-machine gun, 152-3; builds up squad of killers, 153-4; and St Valentine's Day Massacre, 154-5

Cartridges, 13

Casualties, 133, 136, 137-8, 139, 141, 143

Cavalry, in First World War, 128-30; military establishment cling to, 54-6

Cetshwayo (Zulu chief), 82-4

Chauchat gun, 40, 76

Churchill, Winston, 86, 101-2

Carke, I.F., 50-1, 81

Collier's (magazine), 152

Coloney, Myron, 32-3

Colonial warfare, 176

Colonial wars, against Egypt, 85-7, Hausaland, 96-7, Tanganyikans, 88-9, Transvaal, 91, Ugandans, 88, Zulus, 82-4; in Asia, 98, Rhodesia, 89-90; of Seymour Vandeleur, 97-8; role of machine gun in, 101-4

Colonialism, supported by machine guns, 101-4

Colorado, industrial disputes in, 43-4

Colorado Fuel and Iron Company, 43

Colt Company, 83

Cooper Firearms Manufacturing Company, 29

Cripple Creek (USA), 43

Custer, General George Armstrong, 74

de Wiart, Lt-General Carton, 138

Dilke, Sir Charles, 100

Discipline, 105, 171-2

Drill, 124-5

Education, philosophy behind, 104-6

'Fair play', concept of, 106

Farrar-Hockley, A.H., 136

Films, gangster, decline of, 161-2; philosophy of, 160-1; revival of, 162-3; success of, 160

189

Somme, 132-9; as cause of brutalisation, 142-5; advent of Thompson sub-machine gun, 149-52; use in gangster warfare, 152-60; in films, 162-7; advent of tanks spells decline of importance, 167-9; lack of technology held back development, 170-2; destroyed traditional military values, 175-6; lessons to be learned from development, 179-80

Machine gunners, qualities of, 62; lack of, 69; role of defined, 73; African, 95; in First World War, 133; heroism of, 138; experiences described, 142-4

Machine Gun Corps, 114, 122-3, 142-3, 145, 167

Machine tool industry, growth in Europe, 35; growth in USA, 21-5; in 19th century, 173

McLean, James, 32-3

Mahdism, 85-7

Maji-Maji cult, 92-3

Manchester Guardian, 70-1

Marksmanship, 58, 120

Mills, J.D., 25

Millis, Walter, 24

Metallurgy, 173

Meinertzhagen, Colonel Richard, 53, 55

Mons, Angel of, 118-19

Montigny *mitrailleuse*, 63-4

Matabele, wars against, 90-92

Mauser rifle, 126

Maudsley, Henry, 35

Maxim, Hiram, 16, 33-7, 65, 66, 103

Maxim gun, 14, 33-7, 64-5, 69-70, 86, 88-90, 97-8

Maxim-Nordenfelt Company, 14

Muskets, 17

Naval Brigade (British), 84-5

Nelson, 'Baby Face', 159

Newbolt, Sir Henry, 105

Nordenfelt Company, 34

Nordenfelt machine gun, 35

North West Frontier, 98

Norton, General, 72

O'Connor, Chief Detective William. 155-6

Officer class, attitude to machine gun, 16-17; failure to rethink tactics, 17; cut of from social and technical developments, 48; dominated by aristocracy, 48-9; antagonism to progress, 49; belief in personal heroism, 49-51; ignorance of weapon development, 51-3; criticised, 52; adherence to outmoded tactics, 54-7; reaction to machine gun, 56-8; treat machine gun as artillery weapon, 63-4;

conceptions of warfare, 70-1; attitudes drawn from rules of games, 104-5; staffed from public schools, 104; see mass production as threat to them, 175; ignore lessons of colonial wars, 176, 179

Omdurman, battle of, 86-7

Order of American Knights, 28

Ordnance Department (U.S.), 29, 74-5

Palmer (inventor, 1663), 13

Parker, Bonnie, 158-9

Parker, Lieutenance, J.H., 72-3

Philadelphia Public Ledger, 75

Playing the game, 104-6

Poets, and imperialism, 105-6; spiritual confusion of, 144-5

Police (U.S.), and Thompson sub-machine gun, 150; adopt machine guns, 155; marksmanship of, 155; declare war on gangsters, 155-6; battles with gangsters, 157-60

Puckle, James, 13

Public schools, supply members of officer class, 104

Racialism, supported by machine guns, 101-3; theory of, 80

Rhodes, Cecil, 89-90

Ribauldequins, 10-11

Riflemen, 56, 69, 120, 133

Ripley, Colonel, J.W., 25

Romanticism, 118

Roosevelt, Theodore, 73

Royalty, support automatic weapons, 60-1

Russia, 66-8

Russo-Japanese War, 65-8

St Valentine's Day Massacre, 154-5, 161

Sales, 14, 35-42

Sales techniques, 35-7, 38-9

Sammons, James 'Fur', 153-4

Sassoon, Siegfried, 126-7, 144

Savage Arms Company, 41

Scientific American, 155

Seeley, J.E.B., 117

Society, changes to in 19th century, 172-3; structure of, 171

Somme, battle of the, 132-9

Spanish-American War, 72-3

Sport, and warfare, 104-6

Tactical doctrines, 53-5

Tactics, at battle of Ulundi, 84, changes in during American Civil War, 51; in First World War, 114-15; in 1915, 133; in 1918, 167; outmoded, 49-50

Tanks, 167-8

191